ENVIRONMENT
and MAN
Laboratory Manual

TATYANA LOBOVA

Old Dominion University

Kendall Hunt
publishing company

Kendall Hunt
publishing company

www.kendallhunt.com
Send all inquiries to:
4050 Westmark Drive
Dubuque, IA 52004-1840

CONTENTS

ABOUT THIS LAB MANUAL AND NOTEBOOK

This lab manual and notebook was specifically created for the course BIOL 113N *Environment and Man* at Old Dominion University (ODU). The manual provides students with the information necessary to conduct activities and hands-on experiments that will enhance their understanding of the major concepts and issues of the Environmental Science. The content introduces students to the basics of scientific inquiry, scientific method, and hypothesis testing—the universal approaches that scientists use to address the questions about the empirical world. In addition, this volume also serves as a lab notebook.

This Manual consists of three main sections: course introductory information, labs, and appendices.

1. Course introductory material include standard Lab Safety Rules, which are universal for all science labs at Old Dominion University and Lab Policies that are specific for BIOL 113N lab.
2. All labs consist of the activities related to one or more Environmental Science topics. Each lab is generally organized following this format:

Pre-Lab Reading: Selection of topics from the textbook that need to be read prior to the lab in order to better understand and complete lab activities.

Objectives: List of what students are to learn as a result of completing the specific activity.

Activities: List of activities included in the current lab.

Introduction: Brief background information and major terminology on the topic of the current lab; also a purpose and an overview of the exercises.

References: Sources that were used for the preparation of this lab topic.

Hypotheses: Provide a hypothesis or hypotheses being tested in this activity.

Materials: List of materials that are necessary to conduct the activity.

Procedure: Step-by-step description of the activity.

Data: Pages provided to record the raw data and observations for each activity.

Discussion and Conclusion: List of questions pertaining to the completed lab activities that require students to identify potential errors, interpret the collected data, compare the individual data to the group or class data, and draw conclusions related to the original hypotheses. These pages should be completed by students individually, removed from the manual, and submitted to Teaching Assistant (TA) at the end of the lab for completion points.

3. Appendices include additional information on group projects and blank data sheets.

LAB SAFETY RULES

The following policies will be strictly enforced.

- Become thoroughly acquainted with the location and use of safety facilities. Emergency eye-wash stations are located in each laboratory, as an attachment to a faucet on a major sink in each laboratory. Each lab is equipped with a safety shower or has a safety sticker denoting the nearest shower. Use these showers only in emergency and clear the area (lab) when using, as there are currently no floor drains and the floor will be flooded in a matter of minutes. Each laboratory is equipped with a fire extinguisher, which is inspected regularly by the Office of Environmental Health and Safety.

- In the event the **fire alarm** sounds, turn off any burners or hot plates that are in operation and proceed in an orderly manner through the nearest exit and evacuate the building. Do not use the elevator. Do not stand near the loading dock, as this is the triage area and the area where the fire engine will first respond.

- **NO sandals, flip-flops or open-toed shoes, NO shorts** and **absolutely NO** food, drink, or smoking allowed in the laboratory. Be prepared with an old shirt, a lab coat, a pair of sweat pants, and/or sneakers if your dress for the day is inappropriate for a lab setting. Long hair must be tied back. Students with tank tops, bare midriffs, or skimpy shorts will not be admitted to the lab. That means you will be unable to take the quiz or participate in the lab—so you will lose points!

- **All cell phones must be turned off** before entering the lab! Students using cell phone during the lab or leaving the phone on the bench top will be penalized.

- In the event of an accident, the laboratory instructor should be informed immediately. Minor first aid treatment is available at the B.S.S.F. room 207. Please note that the ODU Student Health Center will accept any student with or without insurance coverage on an emergency basis.

- In case of skin contact with an acid or a base, **WASH IMMEDIATELY** for at least 15 minutes. It may seem to be an excessive amount of time but is absolutely necessary to prevent severe blistering and/or burns. Chemicals spilled on clothing should be dealt with appropriately to prevent contact with skin.

- Before beginning an experiment, become familiar with the method of operations and all potential hazards involved including biohazard level, flammability, reactivity, toxicity, and corrosiveness of material. Also be familiar with chemical and biological waste procedures.
- **Notify your instructor immediately** if you are pregnant, color blind, allergic to any chemicals, or have any medical condition (such as diabetes, immunologic defect, etc.) that may require special precautionary measures in the laboratory.
- Upon entering the lab, place all coats, purses, backpacks, etc. on the back of the room and hang on the hangers when possible. Placing of these personal items on the bench top or under the bench is not permitted.
- Report all spills and accidents to your instructor immediately.
- Never leave heat sources unattended.
- All students at a lab table will lose credit for the entire lab if they fail to clean up their lab station before the last student leaves the class.
- All students at a lab table will lose credit for an entire lab, as well as bear the cost of replacement, if any equipment is found missing from that table at the end of class.
- Do not perform experimental work in the laboratory alone.
- Do not perform unauthorized experiments.
- Do not use equipment without instructions.
- All honor code violations (or suspicions thereof) will be reported and pursued to the full extent provided by University policy.
- Individual TA's may have additional policies to which their students must adhere. These will be provided in writing during the first lab meeting.

Additional laboratory policies are included below. All students are responsible for reading, understanding, and following those policies.

LAB POLICIES

LATE ARRIVAL IN LAB

Students arriving late to lab disrupt the TAs and other students in the lab. Students are individually responsible for ensuring that they take whatever steps are necessary to be on time (this includes taking traffic and parking into account). Late arrivals will be dealt with as follows:

▶ **When a quiz is scheduled:**

1. Quizzes will be handed out no more than 5 minutes after the "official" beginning of lab.
2. Students who arrive before the quiz is handed out are considered to be on time and are not penalized.
3. Students who arrive after the quiz is handed out but before the lab lecture begins must remain outside the lab and will NOT be allowed to take the quiz; they will NOT be allowed to make it up. They will be admitted to the lab after the quiz is over and will not otherwise be penalized.
4. Students who arrive after the quiz is over and the lab lecture has begun will NOT BE ADMITTED to the lab. They will not be allowed to turn in assignments (and will receive 0s). They will still be held responsible for the material covered on the quiz (if any) scheduled for the following week.
5. Graduate TAs may modify these guidelines at their discretion under **extraordinary** circumstances (i.e., extraordinary traffic delays vs. the usual slow-downs that happen during bad weather, last-minute family illness, or other emergency, etc.).

▶ **When a quiz is not scheduled:**

1. Students arriving more than 10 minutes after the start of lab will NOT BE ADMITTED to the lab; they are considered absent without excuse. They will not be allowed to turn in their assignments (if any) and will be

penalized 5 points to be deducted from their lab totals at the end of the semester. They will still be held responsible for the material covered on the quiz (if any) scheduled for the following week.

2. Graduate TAs (GTAs) may modify this guideline at their discretion under ***extraordinary*** circumstances (i.e., extraordinary traffic delays vs. the usual slow-downs that happen during bad weather, last-minute family illness, or other emergency, etc.).

STUDENT DISRUPTION DURING LAB

The following policy is based on the Old Dominion University Code of Student Conduct:

1. A student who talks, laughs loudly, or otherwise disrupts a lecture or other lab activity, will receive one clear verbal warning that his/her behavior is disruptive and must stop.
2. If the student persists in those activities, she/he will be penalized 1–5 points depending on the severity of the disruption.
3. If the student still persists in disruptive activity, she/he will be asked to leave the classroom and will receive a score of 0 for all lab activities. If she/he refuses to leave, the GTA will contact campus security to have the student escorted from the room. The student will be written up with an honor code violation in accordance with the University policy.

MISSED LABS

To accommodate students with legitimate reasons for missing lab, we drop your lowest quiz and completion points score prior to calculating your lab score. As with lecture exams, the first item you miss is the one we drop. Thus, even if you miss a lab for a legitimate reason, we will not give makeup quizzes or adjust scores. If you miss more than one lab for legitimate reasons, we will make arrangements on a case-by-case basis. Please review the following policies carefully and contact your GTA if you have questions:

If you miss a lab, the only way you can make it up is to arrange to attend another lab later in the **same week**; we have no mechanism to allow you to makeup an activity after the week in which it is normally conducted. Labs may be made up only if you miss due to a legitimate, documented reason. YOU MUST CONTACT YOUR TA WITHIN 24 HOURS OF YOUR REGULARLY SCHEDULED LAB to make the necessary arrangements unless you can demonstrate that the nature of your emergency prevents you from doing so. If you fail to contact your TA within 24 hours of missing your lab, you will not be allowed to makeup the lab.

You should first attempt to schedule your makeup with your own TA. If you cannot attend his/her lab, you may contact another TA and see if you can attend another lab. You may only attend another lab with the PRIOR PERMISSION of BOTH your own and the other TA. We will work with students with unusual circumstances on a case-by-case basis.

1. If you miss a lab for a legitimate, documented reason and CANNOT make it up, you
   a. may not makeup the quiz (the first quiz missed will be dropped);
   b. may not makeup lab completion question;
   c. may turn in assignments due that day, but only as specified via special arrangement with your GTA. You are responsible for contacting your GTA within 24 hours of the lab and following his/her instructions precisely;
   d. may take the quiz that is given the following week in your regular lab.

2. If you miss lab for a legitimate, documented reason and make it up with another TA (with that TA's prior permission!) you
   a. still need to turn that week's assignment (if any) in to your regular TA in a timely manner (your TA will provide information on how to do this when you notify him/her that you have missed lab);
   b. may take the OTHER TA's quiz during the lab you attend; the other TA will provide your TA with the relevant grades for the day;
   c. may take the OTHER TA's lab completion question;
   d. may take the following week's quiz with your TA in your usual lab.
3. If you miss a lab for a nonlegitimate or undocumented reason and/or fail to notify your TA within 24 hours of missing a lab, you will lose the quiz points for that lab, will be unable to turn in (or receive points for) assignments due that day, and will be penalized an additional 5 points taken from your lab total at the end of the semester.

LAB PENALTIES

Point penalties will be assigned as indicated for each occurrence of the following behaviors:

1. Unexcused absences: 0 points for the lab
2. Failure to bring a lab manual: 5 points
3. Disruptive behavior and/or failure to participate fully in lab activities: 1–5 points
4. Use of the cell phone while in the lab: 5 points
5. **Inappropriate use of lab computers: 10 points for each student in the lab group**

HONOR CODE STATEMENT

The following statement should be typed verbatim on all written work (quizzes, article reviews, projects, etc.) and signed.

In completing this assignment I have abided by the Honor Code of Old Dominion University.

ASSUMPTION OF THE RISK FORM

ODU-ASSOCIATED FIELD PROJECTS

NAME: _Cole Jasso_   CLASS: _Bio 113N_

I agree that as a participant in, or a visitor to, the field projects associated with Old Dominion University and located at _Norfolk main Campus_, I am responsible for my own behavior and well-being. I acknowledge that I have been informed of the general nature of the work conducted at this site, which involves research on live fish, wild animals, plants, and/or work in natural environments, and I understand that it may involve risks to my personal safety beyond those experienced in my everyday activities.

The same elements that contribute to the unique character of this visit can also be causes of loss or damage of personal equipment and accidental injury, illness, or in extreme cases, personal trauma or death. Risks during participation at, or visits to, this field site include, but are not limited to, encounters with venomous snakes, both seen and unseen; encounters with other wild animals; stings by insects such as bees, wasps, and hornets, and other insect; bites by arthropods such as mosquitoes, flies, ticks, and mites; exposure to poisonous or toxic plants including but not limited to poison ivy, poison oak or poison sumac; exposure to arthropod-borne diseases; injury due to attacks by fish, sharks or other aquatic creatures; accidents involving the use of automobiles and watercraft; slipping on wet ground; drowning; injury due to the use of a machete, radio antenna, or other instrument by myself or others; injury due to the activities of fishermen, divers and using hooks, spears, spear guns, etc., including firearms use; vehicular accident; fatigue, exhaustion, heat stress, sunburn and dehydration; becoming lost, disoriented, or separated from my party; or changes in weather conditions. I specifically acknowledge that the activities will occur in nature and that Old Dominion University has little control over natural situations.

I understand that I may be entering remote regions of the field site where communication devices may be ineffective. I also understand that there will be no assurance that special expertise in first aid will be available and, furthermore, that transport to medical care from some areas of this site may be difficult or nonexistent. I understand that in the event of accident or injury, personal judgment must be employed by project personnel regarding what actions must be taken. I understand that it is my responsibility to secure personal health insurance in advance of this visit, if desired, and to take into account my personal health and physical condition.

I further agree to abide by any and all Old Dominion University rules applicable to this trip, and I will take responsibility for abiding by specific requests made of me for my safety, the safety of others, or the welfare of ongoing research projects during the trip. I understand that the University reserves the right to exclude my participation in this field trip if at any time my participation or behavior is deemed detrimental to the safety and welfare of others or to the conduct of ongoing research.

I acknowledge that I have read and fully understand this document. I further acknowledge that I am visiting this field site and accepting these personal risks and conditions of my own free will.

In consideration for being permitted to visit this field site on my own initiative, I hereby release Old Dominion University and its officers, agents, employees, interns, students, and volunteers, from any claim for personal injury or damage to, or loss of, my property which may occur as a result of my participation in this visit or arising out of my participation in this visit, including any such injury, damage or loss that may result from misjudgments made by any officer, employee or agent of the University. I understand that this Assumption of Risk document will remain in effect during any subsequent visits to this field site, unless a specific revocation of this document is filed in writing with the project director, at which time my visits to, or participation at, this field site will cease.

In case an emergency situation arises, please contact ___Danny Jasso___ (name) at ___316-765-4106___ (phone number).

___X___ I represent that I am 18 years of age or older and legally capable of entering into this contract.

_____
Participant's signature

1019 Isabella Dr.
_____
Address

1/23/23
_____
Date

If participant is less than 18 years of age, the following section must be completed:

_____ My child/ward is under 18 years of age and I am hereby providing permission for him/her to participate in this and subsequent field trips and agree to be responsible for his/her behavior during this trip.

_____
Child's name

_____
Parent's or guardian's signature

_____
Address

_____
Date

Lab 1

SCIENTIFIC METHOD

PRE-LAB READING

Textbook: Chapter 1: The Nature of Science.

OBJECTIVES

- Understand how Scientific Method works;
- Demonstrate the ability to use Scientific Method by working through it step by step;
- Gain experience in identifying the experimental variable and control in an experiment;
- Become familiar with the metric system of measurements and with its basic units;
- Gain experience in linear and volumetric measurements, temperature, and mass measurements;
- Practice to plot and interpret graphs.

ACTIVITIES

1. The Scientific Method
2. Metric Measurement

Lab 1

ACTIVITY 1   The Scientific Method

INTRODUCTION

The modern world has been and will continue to be influenced by science. It is an ongoing process that tries to seek order within the natural universe. It is a way to discover new facts as well as expand old ones. Science comes from the Latin noun "for knowledge." It can be broken down into many different disciplines. A few are Biology, Chemistry, and Physics. Biologists study the origin and life history of plants and animals, as well as such topics as Ecology, Zoology, and Botany. Chemists study the nature and composition of substances, and the laws which govern their relations. Physicists study matter and energy, including mechanics, heat, light, and electricity. Even though these scientists focus on different areas, they all acquire knowledge by a specific method—**The Scientific Method.**

▶ The Scientific Method consists of four main steps.

The first is a primary observation
At this time, one recognizes and states a problem. Sometimes the problem emerges by chance or luck. A literature search is usually done to find any information on the problem that has been observed.

> EXAMPLE: Casual field observation indicated that Zebra mussel colonization was markedly different between the northern (U.S.) side and the southern (Canadian) sides of Lake St. Clair. Information was obtained by doing a literature search. Zebra mussels were probably introduced into Lake St. Clair from ship ballast in late 1985 or early 1986. Since then, they have spread rapidly into all five of the Great Lakes. The information that was found was from European studies. So, the potential impact of this species on the Great Lakes is not yet known.

The second step is to create a hypothesis
A scientist tries to think of possible explanations for the problem at hand. A hypothesis should be testable, that is, generate definite predictions and be consistent with all well-established facts. It must also be falsifiable.

> EXAMPLE: The hypothesis for the above observation states that there were no differences in growth rates between mussels located at the northern side and those at the southern side of Lake St. Clair.

The third step is to design and execute an experiment to test the hypothesis
The experiment must be accurate, repeatable, and objectively based, not subjectively based. Objective (quantitative) experiments allow the study to be tested statistically. There are two groups involved in setting

up an experiment. One being the experimental variable group and the other being the control group. The control group has the same conditions except for the experimental variable. The experiment is better supported by the collection of more data. Performing the experiment only once is not enough. The results received from that one time may be a fluke (just due to chance).

EXAMPLE: Six cages of 50 Zebra mussels (3–5 mm starting length) were set out at two sites, one on the U.S. side and the other on the Canadian side of Lake St. Clair. There were placed at a depth of 3–4 m. Three of the six cages were monitored bimonthly for one year for shell length, height, and width, as well as live weight. The remaining three cages were measured only two times over the year and used as a control to evaluate the effect that more frequent sampling (measuring) had on growth. At the end of one year's sampling period, the Zebra mussels were taken out and the cages restocked with a new group of animals. The study was run the same way as the previous year. Thus, the study run was for a two full-year period.

The fourth step occurs after a hypothesis has been supported by a great amount of testing

It then becomes a theory. A theory should relate previous facts that appeared to be unrelated, as well as grow as more facts are found. A theory is tested and retested, and it is only valid if it answers every case. At this point in time, a scientist may wish to publish the findings in a journal so that other scientists can learn from the studies.

Lab 2
BIODEGRADATION PROJECT SET UP

PRE-LAB READING

Textbook: Chapter 22: Managing out Waste.

ACTIVITIES:

Project Set Up

Lab 2

ACTIVITY 1 Biodegradation Project Set Up

See Appendix I: Biodegradation Project and Lab Report: Activity 1: Project Set Up.

Lab 3

TOXICOLOGY

PRE-LAB READING

Textbook: Chapter 14: Environmental Health and Toxicology.

ACTIVITIES

Abiotic Influence on *Daphnia*.

INTRODUCTION

An organism must interact with the nonliving or abiotic environment, no matter where it lives. Abiotic factors can be of physical or chemical types. They include such things as sunlight, temperature, pH, humidity, and gases found both in the atmosphere and in the soil.

Not all organisms are affected the same way by the same abiotic factor. For example, the amount of sunlight that hits a desert in North America may be too much for one organism but just right for another. Organisms dealing with extreme heat of the desert may limit their activities to only the early mornings or late evenings. Others may hibernate or estivate until the conditions are more favorable, that is, until the rainy season occurs. So, it is not enough for the organism to just interact with the abiotic factors. Organisms must be able to tolerate those factors. If they cannot tolerate the factors for one reason or another, they must move out of the area in search of more tolerable conditions. However, if that organism is a plant of some sort it cannot move and may eventually die. In time, all the plants of that kind will either genetically adapt to tolerate the factor or die off from that area and be replaced by plants that can better tolerate the abiotic factor in question.

Animals have it a little easier than plants in that they can move out of an area of low tolerance to another area of high tolerance. However, these animals that move may be moving into areas with new abiotic factors. This results in the animal requiring a new set of tolerance levels.

Some abiotic factors are put into the environment by human activities. These factors may cause a greater threat to organisms because they have not had the chance to develop any tolerance to these factors.

A major problem is acid rain. Human activities, especially motor vehicles' emissions, increase the amount of sulfur dioxide (SO_2) and nitrogen oxides (NO) released into the atmosphere. Once in the atmosphere, they are circulated with the air currents. Chemical reactions take place in the atmosphere changing SO_2 and NO into weak forms of sulfuric acid (H_2SO_4) and nitric acid (HNO_3). As a result, the acid returns to the Earth in the form of rain. This rain decreases the pH of the areas that the rain falls on. Effects are seen in both aquatic and terrestrial organisms. In many cases, the tolerance levels of these organisms to acid rain are unknown. In this laboratory experiment, the tolerance levels of *Daphnia*, the freshwater flea, to four chemicals will be tested.

OBJECTIVES

1. Describe several abiotic factors.
2. Explain how abiotic factors have different effects on organisms.
3. Describe effects that may occur due to varying abiotic factors.
4. State several ways an organism can overcome the effects of abiotic factors.

MATERIALS

For the purpose of this project, the following materials need to be acquired and utilized:

- 5 Petri dish: 4 with four sections and 1 standard
- 1%, 0.1%, 0.01%, and 0.001% nitric acid (HNO_3)
- 1%, 0.1%, 0.01% and 0.001% hydrochloric acid (HCl)
- 1%, 0.1%, 0.01% and 0.001% sulfuric acid (H_2SO_4)
- 1%, 0.1%, 0.01% and 0.001% ammonia (NH_3)
- Distilled water
- 80+ *Daphnia*
- Dissecting scope
- Marker
- Droppers

Abiotic Influence on *Daphnia*

PROCEDURES

▶ STEP 1

1. Obtain 5 Petri dishes; four of these should be the types that are divided into four sections and the fifth should be standard.
2. For nitric acid (HNO_3) solution, label each section of the divided Petri dish as follows: 1%, 0.01%, 0.001%, and 1%. Repeat the procedure for the other solutions (HCL, H_2SO_4, and NH_3) also. Place distilled water in the nondivided Petri dish.
3. Using the graduated cylinder, measure out 5mL of 1% HNO_3 and pour into the dish section labeled "HNO_3 1%." Repeat until all the dishes are filled with the appropriate amounts of the respective solutions.

NOTE: After each use of the graduated cylinder, rinse it out with distilled water.

---

4. Add distilled water to the nondivided Petri dish.
5. Take a container containing *Daphnia*.
6. Using a dropper, count out 5+ *Daphnia* and transfer them to each section of the divided Petri dish. (Transfer the same number to each section of the Petri dish.)
7. Repeat the above process until *Daphnia* are transferred to the various acid and base solutions.
8. Also transfer 5+ *Daphnia* to the distilled water Petri dish.

▶ STEP 2

1. After 30 minutes observe the number of *Daphnia* that are alive. Record your information in the table provided.
2. Measure the pH of each solution. Record in the table provided.
3. Count and record the number of living *Daphnia* that is present in each Petri dish.
4. Look at each dish under the dissecting microscope or transfer the *Daphnia* to a slide and use a regular microscope.

RESULTS

Dish No.	Treatment	pH Reading	No. Live *Daphnia*
1	Distilled water	7	100
2	1% HNO_3	5.7	0
3	0.1% HNO_3	6	4
4	0.01% HNO_3	6.4	27
5	0.001% HNO_3	6.6	68
6	1% HCl	5.8	0
7	0.1% HCl	5.9	4
8	0.01% HCl	6.4	30
9	0.001% HCl	6.7	73
10	1% H_2SO_4	5.2	0
11	0.1% H_2SO_4	5.6	0
12	0.01% H_2SO_4	5.9	2
13	0.001% H_2SO_4	6	6
14	1% NH_3	9.7	0
15	0.1% NH_3	9.4	0
16	0.01% NH_3	9	1
17	0.001% NH_3	8.4	22

Lab 4
WATER POLLUTION

PRE-LAB READING

Textbook: Chapter 15: Freshwater Systems and Resources.

ACTIVITIES

Water Quality Analysis.

Lab 4

ACTIVITY 1 Water Quality Analysis

OBJECTIVES

After completing this experiment, you should be able to:

■ Be familiar with local water quality reports and know how to read them
■ Perform basic water quality tests for select pollutants
■ Understand how the capacity of water to hold oxygen changes as a result of temperature and pollution

INTRODUCTION

Water is continuously renewed and recycled through the **hydrologic cycle** (**Fig. 4.1**). It moves through biotic and abiotic components of the environment as a liquid, a solid and a gas. Liquid water from Earth's surface moves into the atmosphere via **evaporation** and **transpiration**, the release of water as a vapor through the leaves of plants [1]. Once in the atmosphere as a vapor, the water mixes with components of the atmosphere and is carried by air currents until it is deposited back on Earth as **precipitation**. The rain, snow or ice deposited on Earth can be taken up by plants, used by animals or allowed to reach surface water bodies directly or as **runoff**. In addition, some precipitation along with some surface water infiltrates the soil and rock to recharge underground aquifers. Ground water moves very slowly but eventually can reach the surface again by means of natural springs and human-drilled wells. The average age of ground water is estimated to be about 1,400 years and some may be thousands of years old [1].

Although water may seem to us to be in abundant supply, globally it is a very limited resource [1]. Approximately 75% of Earth's surface is covered in water. If this water was evenly distributed across the planet, it would completely cover the surface of the Earth to a depth of three kilometers (1.9 miles) [2]. Despite the apparent abundance, less than 1% of the water on Earth is readily available to people for drinking and irrigation. Most of the water (97.5%) is in the oceans and is therefore too salty for human consumption or irrigation. Of the remaining 2.5% that comprises freshwater, 79% is frozen in glaciers and ice caps, 20% is in underground aquifers and only 1% is on the surface of the Earth [1]. Most of the surface water is in the form of lakes (52%), followed by soil moisture content (38%), water vapor in the atmosphere (8%), rivers (1%) and water within living organisms (1%) [1].

Water is not equally distributed across the Earth and people are not distributed across the Earth in accordance with water availability. In fact, many areas that have high population densities are poor in water resources. This problem leads to inequities in per capita water resources among and within nations [1]. Freshwater is a renewable resource. However, if we use water faster than it can be replenished, then we have effectively reduced the amount of water available per person or per civilization that can be used over a lifetime. At present, much of the world's freshwater consumption is unsustainable. "At least 1.7 billion people live in areas of water scarcity, that is, areas where water availability per person is less than 1,000 m^3 (35,000 ft^3) per year. The number of people facing water scarcity is expected to grow to at least 2.4 billion by 2025" [1].

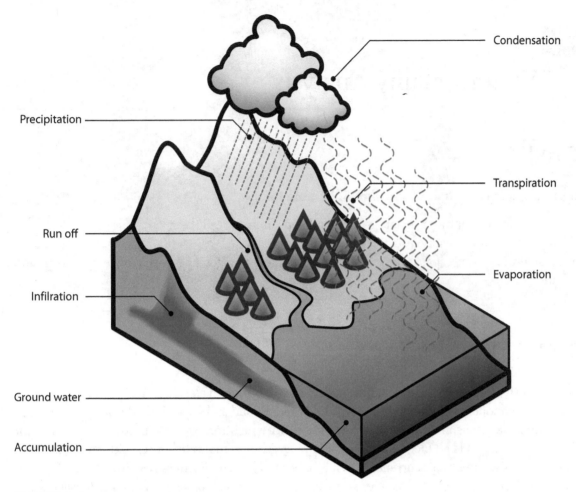

FIGURE 4.1.
Illustration of the hydrologic cycle.

Along with freshwater quantity and distribution issues is the concern for water quality. In 1999, the World Commission on Water concluded that over half of the world's major rivers are seriously polluted to the point of poisoning the surrounding ecosystems and threatening the health and livelihood of the people who depend on them [1]. Water pollution can come from both point and non-point sources. **Point sources** of pollution come from a single location such as a municipal waste water treatment plant discharge pipe. **Non-point sources** of pollution come from multiple sites over large areas such as pesticides used on residential lawns, fertilizer runoff from farms and oil from city streets. Regardless of the source of pollution, **surface water pollution comprises three broad categories**:

1. **Biological Pollution**—Comes from disease-causing organisms that live in surface waters [1].
2. **Chemical Pollution**
   a. *Natural Chemical Pollution*—Commonly caused by phosphorus and nitrogen. Rain and/or irrigation cause runoff of excess nutrients (nutrients not absorbed by plants) from farmland, golf courses, lawns and sewage into nearby water bodies.
   b. *Artificial Chemical Pollution*—Pesticides, petroleum products, pharmaceuticals and other synthetic chemicals that end up in our waterways. This category also includes inorganic pollutants that enter the environment as a direct result of human activities (examples are metals such as arsenic and lead). Also includes acids from precipitation and mine drainage [1].

3. Physical Pollution

   a. *Sediment Pollution*—Sediment carried by rivers can be a pollutant. Some sediment in rivers is normal, but increases brought on by mining, clear-cutting, real estate development, and poor farming practices are problematic [1].

   b. *Thermal Pollution*—Defined as an unnatural change in a water body's temperature that impacts the amount of dissolved oxygen the water body can hold. Warm water hold less oxygen and cold water holds more oxygen. Water is often taken from rivers or lakes to cool industrial facilities. The water is heated during the process then returned to the surface water body where it raises the overall temperature of the water. Removal of vegetation along streams raises water temperature whereas damming of rivers lowers water temperature [1].

   c. *Solid Waste Pollution*—Solids disposed of by humans that end up in our waterways. E.g. tires, fishing nets, plastic bottles etc.

Groundwater pollution is considered to be a more serious problem than surface water pollution because it is longer lasting. Chemicals are broken down at a much slower rate since groundwater has less dissolved oxygen, microbes, minerals and organic matter [1]. In the United States, we pump 60% or about nine billion gallons of our liquid hazardous wastes underground each year, some of which have found their way into drinking water supplies [1]. Other sources of groundwater pollution include: leaking underground storage tanks that contain septic materials, industrial chemicals and oil and gas; agricultural practices that leach pesticides and nitrates from fertilizers; pathogens that enter through improperly maintained wells; toxic chemicals from manufacturing industries; and radioactive wastes and other chemicals from military sites [1].

In 1974, Congress passed the **Safe Drinking Water Act** to protect public health by regulating public drinking water supplies [2]. Today the U.S. Environmental Protection Agency (EPA) sets standards for about 90 contaminants in drinking water [3]. The EPA sets two types of standards: primary and secondary. **Primary standards** are legally enforceable and aim to protect public health by limiting the level of the contaminant in drinking water. **Secondary standards** are not legally enforceable and typically involve contaminants that may cause cosmetic effects (skin or tooth discoloration) or aesthetic effects (taste, odor or color) in drinking water [4]. Most municipalities provide their consumers with annual water quality reports. The most current water quality report for the City of Norfolk can be found online at:

http://www.norfolk.gov, search: water quality report.

There are a few terms you need to be familiar with in order to read and understand the report.

Maximum contaminant level or **MCL** is: "the highest level of a contaminant that is allowed in drinking water" [4]. The MCL is the enforceable standard.

Maximum contaminant level goal or **MCLG** is: "the level of a contaminant in drinking water below which there is no known or expected health risk" [4]. In other words, the MCLG is the safe level for a contaminant. The MCLG is not an enforceable standard.

Typically, the MCL is set as close as possible to the MCLG, but for some contaminants the MCL or the enforceable standard is higher than the MCLG or the safe level of a contaminant. This typically occurs when complete removal of a contaminant would be too cost prohibitive or when the removal of a contaminant would be a trade-off between two different health issues. For example, we disinfect drinking water to remove bacteria, viruses and other pathogens that cause diseases in humans. However, the disinfection process creates chemical byproducts that are known to cause cancer. We knowingly allow low levels of the disinfection byproducts that cause cancer to be in our drinking water so that we can prevent illnesses associated with water borne pathogens. This leads us to our last two terms:

Maximum residual disinfectant level or **MRDL** is: "the highest level of a disinfectant allowed in drinking water."

Maximum residual disinfectant level goal or **MRDLG** is: "the level of a drinking water disinfectant below which there is no known or expected risk to health" [4].

The purpose of this lab exercise is to qualitatively assess two water samples for common water contaminants. In addition, we will look at differences in oxygen levels as a result of temperature and fertilizer inputs. The following is a list of the contaminants you will be testing for in your samples along with information for why the EPA has set a standard for the particular substance.

Turbidity is a measure of water's cloudiness [4]. It has no health effects but it can interfere with the disinfection process. Turbidity can provide a medium for microbial growth, and turbid waters may indicate the presence of disease causing organisms [4]. Turbidity cannot exceed 1 NTU in drinking water (NTU is a unit of measurement = nepholometric turbidity unit) [4]. NTU can be converted to cm of H_2O using standardized conversion tables available online. E.g., 1 NTU = >60 cm H_2O.

Nitrates enter drinking water supplies as runoff from fertilizer, as leachate from septic tanks or sewage and from erosion of natural deposits. Nitrates can cause serious illnesses or death in infants below the age of six months. Both the MCL and the MCLG for nitrates is 10 mg/L in drinking water [4].

Iron enters drinking water from the erosion of natural deposits. Iron is generally not toxic to humans but high levels can give water a bitter taste. In addition, iron can stain laundry, porcelain, and enamel brown [5]. Iron has a secondary drinking water standard of 0.3 mg/L [4].

Phosphorus is not currently regulated by the EPA. Phosphorus enters drinking water supplies from erosion of natural deposits. Human sources of phosphorus include fertilizers, detergents and sewage. As mentioned previously, phosphorus is one of the key substances that cause nutrient pollution which leads to eutrophication and hypoxic conditions.

Copper is a heavy metal that enters drinking water supplies from the erosion of natural deposits and the corrosion of household plumbing systems. Copper is toxic to fish at low levels and short term exposure causes gastrointestinal distress in people. Copper is regulated by treatment techniques that require systems to control the corrosiveness of their water. If the concentration of copper exceeds 1.3 mg/L, drinking water plants must perform additional treatment steps until the level of copper is reduced. The MCLG for copper is 1.3 mg/L [4].

Ammonia is not currently regulated by the EPA for drinking water supplies. Ammonia is a decomposition product of materials containing nitrogen (sewage, fertilizer runoff, manure from feedlots). High concentrations of ammonia are toxic to fish and other aquatic organisms. Ammonia toxicity varies but increases with increasing pH and water temperature. The EPA recommends that ammonia concentrations do not exceed 0.02 mg/L in freshwater to protect aquatic life [5].

Silica is not currently regulated by the EPA for drinking water supplies. Silica is a common constituent of igneous rocks, quartz and sand and enters water through the erosion of natural deposits [5]. Silica is an important, desirable addition to water. It is the principle component of diatom cell walls. Diatoms are a type of algae and often make up the bottom layer of the food pyramid in lake communities.

Sulfide is not currently regulated by the EPA for drinking water supplies. Sulfide is a common water pollutant associated with oil fields, acid mine drainage and areas where bituminous coal is mined [5].

Chlorine is a disinfectant that is used to control microbes in drinking water supplies. Chlorine can cause eye and nose irritation as well as stomach discomfort if ingested with water. Both the MRDLG and the MRDL are 4 mg/L in drinking water [4].

Cyanide enters drinking water supplies in discharge from steel and metal factories, plastic manufacturers, and fertilizer factories. If ingested with water, cyanide can cause nerve damage and thyroid problems. Both the MCL and the MCLG are set at 0.2 mg/L [4].

Chromium is a heavy metal that enters drinking water supplies through the erosion of natural deposits. Chromium also enters water supplies when it is discharged from steel and pulp mills. Chromium can cause allergic dermatitis in individuals who drink contaminated water. Both the MCL and the MCLG for chromium is 0.1 mg/L [4].

pH ranges from 0 to 14 with 7 being neutral. Values greater than 7 are alkaline and values less than 7 are acidic. The pH is an important aspect of water quality. Changes in pH may increase or decrease the amount and toxicity of many pollutants, especially heavy metals. Natural rain water has a pH around 5.6. Most natural water bodies have a pH between 6 and 8.

MATERIALS

◼ Water Pollution Test Kit

METHODS

You will be given two water samples (pond water and tap water) to analyze for basic water quality parameters.

▶ Determination of Water Quality Parameters in Pond and Tap Water

Next follow the directions below to evaluate the water quality parameters. **NOTE: Each group only performs 1 water quality test.** You will not have time to test for all parameters. The class will share results for each of the tests.

Each group of two students performs the water quality test on both pond and tap water. To confirm results, each test should be repeated at least once on each water sample (a total of 2 or 3 tests per water sample).

These tests are colorimetric, qualitative assessments and will only indicate the presence or absence of each parameter. If the parameter is present, as indicated by the color change, record with a plus (+) symbol. If the parameter is absent, record with a negative (–) symbol. Record your results in the appropriate section of the Lab Report.

▶ Procedures for Detection of Common Water Pollutants

Ammonia Nitrogen

1. Measure a 10 mL water sample into a calibrated tube.
2. Add 1 drop of Ammonia Test Solution #1 to the water sample and mix.
3. Add 8 drops of Ammonia Test Solution #2 to the water sample and mix.
4. If ammonia nitrogen is present in the sample, a yellow color will develop. Allow 8 to 10 minutes for full color development.

Chlorine

1. Fill a Chemplate® cavity approximately 2/3 full with the water to be tested.
2. Add 2 drops of Chlorine Test Solution and mix with the plastic spatula.
3. If chlorine is present, a yellow color will develop. Allow 5 minutes for full color development.

Chromium (Chromate)

1. Measure a 10 mL water sample in a calibrated tube.
2. Add 2 drops of Chromium Extracting Solution to the water sample. Place the cap on the calibrated tube and shake.
3. Add a level spatula of the Chromate Indicator Powder. Replace the cap and mix the sample until the powder is dissolved.
4. A reddish-purple color forms in the presence of chromate and the amount of color directly proportional to the amount of chromium (chromate) present in the sample.

Copper

1. Fill a Chemplate® cavity approximately 2/3 full with the water to be tested.
2. Add 1 drop of Copper Test Solution #1. Mix and allow to stand for 1 minute.
3. Add 2 or 3 drops of Copper Test Solution #2. Mix and allow to stand at least 2 minutes but not more than 10 minutes.
4. An orange-colored solution indicates the presence of copper.

Cyanide

1. Measure a 10 mL water sample into the calibrated tube.
2. Add 2 drops of Cyanide Test Solution #1 and mix.
3. Add 2 drops of Cyanide Test Solution #2 and mix.
4. If cyanide is present, a pink color will develop that turns violet in a few minutes. Allow approximately 10 minutes for the color to develop.

Iron

1. Measure a 5 mL water sample in the calibrated tube.
2. Add 5 drops of Iron Test Solution #1 (this solution is 5% Sulfuric Acid—be careful!)
3. Add 1 level spatula of Iron Indicator Powder to the sample. Replace the cap and mix to dissolve.
4. If iron is present, a wine red color will develop. Allow 2 minutes for full color development.

Nitrate/ Nitrogen

1. Place approximately 3 mL of water sample in the calibrated tube.
2. Add enough Nitrate Test Solution #1 (2 mL) to bring the sample up to 5 mL and mix well.
3. With the plastic spatula, add 2 level measures of Nitrate Indicator #2 Powder.
4. Replace the cover and shake until the powder is completely dissolved.
5. If nitrate nitrogen is present, a very light pink color will develop (trace amounts). A reddish purple color will develop with high concentrations of nitrate nitrogen. Allow 5 minutes for full color development.

Phosphorous (Phosphates)

1. Measure a 5 mL water sample in the graduated tube.
2. Add 15 drops of Phosphate Test Solution #1 and mix. Allow to stand 3 to 5 minutes. A light yellow indicator may appear.
3. Add 2 or 3 drops of Phosphate Test Solution #2. Replace the tube cover and mix.
4. If phosphate is present, a blue color will form immediately.

Silica

1. Measure a 5 mL water sample in the calibrated tube.
2. Add 3 drops of Silica Test Solution #1 and mix.
3. Add 6 drops of Silica Test Solution #2 and mix.
4. Add 4 drops of Silica Test Solution #3 and mix.
5. Add 1 drop of Silica Test Solution #4 and mix.
6. If silica is present, a blue color will form immediately.

Sulfide

1. Measure a 5 mL water sample in the graduated tube.
2. Add 15 drops of Sulfide Test Solution #1 and mix. (Note: This solution has a high sulfuric acid content and care should be taken!)
3. Add 3 drops of Sulfide Test Solution #2. Mix and allow to stand for 1 minute.
4. Add 20 drops of Sulfide Test Solution #3 and mix.
5. If sulfide is present, a blue color will appear.

pH

1. Place a small sample of the water to be tested (8 to 10 drops) in a cavity of a Chemplate®.
2. Add 1 drop of Universal pH Indicator and mix with the plastic spatula. Compare the color that immediately appears with the list below.

pH 1	Cherry Red	pH 6	Yellow
pH 2	Rose	pH 7	Yellow-Green
pH 3	Red-Orange	pH 8	Green
pH 4	Orange-Red	pH 9	Blue-Green
pH 5	Orange	pH 10	Blue

REFERENCES

Scott, B., and J. Withgott. 2005. *Environment: The Science Behind the Stories.* Pearson Education, Inc., publishing as Benjamin Cummings.

USEPA 1999. Understanding the Safe Drinking Water Act. EPA 810-F-99-008.

USEPA. Drinking Water Contaminants. http://water.epa.gov/drink/contaminants/. Last updated May 4, 2011. Last accessed on June 9, 2011.

USEPA. List of Drinking Water Contaminants and MCLs. http://water.epa.gov/drink/contaminants/#List. Last updated May 4, 2011. Last accessed on June 9, 2011.

Franks, J. 1998. *Discovering the Balance: Environmental Science Lab Manual.* New York: The McGraw-Hill Companies, Inc., Primis Custom Publishing.

Qualitative Introduction to Water Pollution Kit. Cat. No. 19. LAB-AIDS®, Inc. 17 Colt Court, Ronkonkoma, NY, 11779.

Dissolved Oxygen. Water Quality Test Kit. Instruction Manual. Cod 7414/5860.LaMotte Company. PO Box 329. Chestertown, MD, 21620.

Lab 5

RESOURCE CONSUMPTION

PRE-LAB READING

Textbook: Chapter 16: Marine and Coastal Systems and Resources.

ACTIVITIES

Marine Fisheries.

Lab 5

ACTIVITY 1 Marine Fisheries

OBJECTIVES

After completing this experiment, you should:

■ Understand the impact of humans on a global scale
■ Understand the concepts of carrying capacity and "tragedy of the commons"
■ Be familiarized, through the provided example, with the real threat confronting global marine fisheries

INTRODUCTION

Though we may be inclined to believe otherwise, natural resources are not unlimited. Every species, including man, has a limit to the number of individuals that may be supported by a particular environment. This condition is known as **carrying capacity** and is defined as "the maximal population size of a given species that an area can support without reducing its ability to support the same species in the future." More specifically, it is "a measure of the amount of renewable resources in the environment in units of the number of organisms these resources can support" [1]. Basically, a larger region or a region with a greater abundance of natural resources will be able to support more individuals. Likewise, a given area will be able to support more individuals having lower energy requirements than those requiring a greater consumption of resources [1].

Carrying capacity for a given region is not set in stone. The number of people an area can support may increase with improved technology, or lessen by degradation of its natural resources. With increased usage of global natural resources comes a responsibility to preserve a quality of life for later generations. The idea that we consider future generations in our utilization of natural resources is described by the term **sustainability**. Sustainability is defined as "the ability to meet humanity's current needs without compromising the ability of future generations to meet their needs" [2].

The example of resource consumption we will examine in this laboratory exercise concerns global marine fisheries. In a seminal paper published in 1968, Garrett Hardin discussed the concept of the "tragedy of the commons." This concept is developed around the idea of a "commons," a free resource available to all. If used sparingly, and in a communal manner, the resource may last indefinitely. However, if one group decides to take advantage of this free resource and thereby uses more than it is due, the resource is rapidly degraded and is ultimately lost to everyone [3].

Hardin uses marine fisheries as a prime example of a global commons. All coastal nations have free access to the world's fisheries. While national claims to coastal waters vary in distance, offshore waters are open to all. They are not, however, an inexhaustible resource. Because of this, international treaties are designed to institute quotas on the various fisheries. If quotas are equitable and adhered to, marine resources may last a long time. However, if instead a country disregards treaties and the global good, over-exploiting this resource, the fisheries will rapidly trend toward depletion [3].

Global fisheries are under more pressure than at any other time in history. From 1950 to 1990, there was an estimated four to fivefold increase in world annual fish catch. The average annual fish consumption is approximately 59 pounds per person in industrialized countries, or almost three times the consumption (20 pounds/person) of developing countries. It is estimated that 70 percent of the world's marine fisheries are fully exploited or over-exploited.

The number of people fishing and practicing aquaculture has doubled in the last 30 years. Technology used to catch fish and the number of fish caught varies widely. Modern fleets are the most environmentally destructive, utilizing advanced technology to locate large schools of fish. Once the fish are located, these fleets use very effective and often destructive techniques to optimize their take. Modern fishing technology is designed to catch many fish quickly, with little regard for deleterious consequences. Coral bed destruction, catching of non-target species and an extensive wasted by-catch—around 27 million tons annually—are some of the results of this type of fishing.

Increasingly, **aquaculture** or "fish-farming" is being utilized as an alternative to open ocean fishing. Almost a third of all fish used for food is produced via aquaculture. While this type of "farming" may protect wild fish stocks, fish farming may have a negative impact on the environment. For instance, mangrove forests are being destroyed to make room for shrimp farms, wild fish stocks are being depleted to be used as food for farmed fish, and extensive waste is being generated from caged or farmed fish. Currently, there is extensive research conducted on ways to limit or alleviate the negative environmental impacts of aquaculture. This initiative is important as we may have to rely more (or equally) on aquaculture than open ocean fishing to sustain a continuously growing human population. Countries with emerging economies will rely more on a meat diet and thus the demand for seafood will increase along with growing economies/populations. Aquaculture thus appears to provide a useful addition to open ocean fishing that not only protects wild fish stocks from being depleted, but also contributes to an efficient production of a high quality food source. As aquaculture will likely never replace open ocean fishing, sustainable fishing practices are seen as a viable option for preserving the world's fish stocks. Like other sustainable practices, sustainable fishing is dependent on effective incentives, generally monetary, being provided to people making their living from the oceans. Prevention of a "tragedy of the commons" at sea can only be prevented if the initiative to do otherwise is efficiently provided on a global scale [4].

MATERIALS

- Dice
- Two types of beans (brown and white)
- Plastic cups

METHODS

This experiment simulates effects of fishing on populations of marine fish. You will be working with three scenarios.

Scenario 1: Represents traditional fishing techniques, which may be for the most part, sustainable.

REFERENCES

Daily, G. C., and P. R. Ehrlich. 1992. "Population, Sustainability, and Earth's Carrying Capacity." *BioScience* 42: 761–771.

Berg., L. R., and M. C. Hager. 2007. *Visualizing Environmental Science,* 12. Hoboken, NJ, USA: Wiley and Sons.

Hardin, G. 1968. "The Tragedy of the Commons." *Science* 162: 1243–1248.

Fishing for the Future. Fishery Facts. http://www.pbs.org.emptyoceans/educators/activities/fishing-for-the-future. html. 2002–2011. © by Facing the Future: People and the Planet, 2004.

Lab 6

AIR POLLUTION

PRE-LAB READING

Textbook: Chapter 17: Atmospheric Science, Air Quality, and Pollution Control.

ACTIVITIES

1. Biodegradation Project Observation
2. Estimating Ozone and UV Levels

Lab 6

ACTIVITY 1   Biodegradation Project Observation

Perform Biodegradation Project Observation following the instructions in the Appendix I.

ACTIVITY 2   Estimating Ozone and UV Levels

OBJECTIVES

- Explain the sources of air pollution
- Define the Air Quality Index
- Explain the sources of ozone and its effect on human health
- Test and evaluate local ozone level
- Determine the UV light intensity

MATERIALS

- Thermometer
- Ozone test strip
- UV indicator card
- Transparent tape
- Timing device

INTRODUCTION

▶ Combustion Engines

Modern automobile engines work by combustion, the burning of fuel. Inside the cylinders of the engine, fuel is mixed with air at high pressure. A spark inside the pressurized cylinder ignites the fuel. The energy from this controlled explosion is converted to mechanical energy that turns a crankshaft and operates the drive wheels of the vehicle. When fuel is burned, water vapor, carbon dioxide, and various other chemical products result, including carbon monoxide and hydrocarbons, both of which are hazardous pollutants. Depending on characteristics of the fuel and the efficiency of burning, different engines produce exhaust with different composition. To help reduce air pollution, federal and state laws regulate the amount of emissions allowed by various engines.

▶ Greenhouse Gases and Climate Change

Various climate data suggest that over the past few decades earth has experienced a period of rapid warming. These data have prompted scientists to explore the extent to which industrialization has influenced current changes in global climate. Mounting evidence links increasing levels of carbon dioxide (a green-house gas) in the atmosphere and higher average global temperatures. Atmospheric concentrations of the greenhouse gases carbon dioxide (CO_2), nitrous oxide (N_2O), and methane (CH_4) have been increasing for decades. When solar energy is absorbed by the earth, it is degraded and reemitted as thermal infrared radiation. Most of this escapes into space. Because of their structure, molecules of greenhouse gases absorb some of this radiation, become warmer, and radiate heat. Some of this heat radiates back to earth, some is absorbed by other greenhouse gas molecules, and some continues out to space. When the atmosphere contains more greenhouse gas molecules, less radiant heat is lost to space. Although a certain quantity of greenhouse gas is essential for retaining heat and making the earth warm enough for life as we know it, a large increase in these gases may alter the earth's climate.

▶ Carbon Dioxide

Atmospheric carbon dioxide levels have varied greatly throughout earth's history, with higher levels often corresponding to warmer periods. Measurements of atmospheric CO_2 have indicated a steady increase in the last 200 years. Natural sources make up the majority of CO_2 emissions and include ocean release, animal and plant respiration, organic matter decomposition, forest fires, and volcanic eruptions. Although human activity accounts for only a small percentage of CO_2 emissions, human output of CO_2 has grown steadily since industrialization. This large increase over a short period may have outpaced the natural carbon cycle's capacity for moderating the atmospheric level, resulting in a slight but steady rise in atmospheric CO_2. The major contributors of anthropogenic CO_2 are the burning of fossil fuels (oil, coal, and natural gas) and the destruction of forests (which take carbon dioxide from the atmosphere during photosynthesis).

▶ Smog Formation

One type of pollution resulting from fossil fuel combustion is photochemical smog. This grey-brown haze often forms over cities due to a reaction involving primary and secondary pollutants, ultraviolet (UV) light, and heat.

     Primary pollutants such as nitrogen oxides (NOx) and volatile organic compounds (VOCs) are released directly into the atmosphere. These pollutants react and form secondary pollutants including ozone, organic nitrates, oxidized hydrocarbons, and photochemical aerosols. It is important to remember that ozone occurs in two layers of the atmosphere. The "good" ozone layer exists in the stratosphere between 10 and 30 miles from the earth's surface. It protects organisms on earth from the sun's harmful UV rays. Any decrease in the naturally occurring ozone in the upper atmosphere allows more UV radiation to penetrate. In the past, this protective ozone layer was reduced by human activities, mainly chlorofluorocarbons that were used as refrigerants for many years. There is evidence that the layer has been recovering since those chemicals were replaced by alternatives. In contrast, the "bad" or "ground-level" ozone in photochemical smog exists in the atmospheric layer closest to earth, the troposphere. When inhaled, ozone can decrease lung capacity and aggravate asthma. Ground-level ozone is created when NOx and VOCs, released mainly by vehicles and industries, react in sunlight. The VOCs include hydrocarbons that result from inefficient combustion. VOCs also enter the atmosphere directly through evaporation (e.g., gasoline fumes). Smog formation is affected by environmental factors such as wind, temperature, ultraviolet radiation, and topography. For example, smog is very likely to develop during a hot, sunny period in a heavily populated valley that is pro-tected from wind. High temperatures and more intense UV radiation tend to increase smog formation. The extent of cloud cover, the time of day, and the season all affect the intensity of UV radiation. The effects of smog on human health vary. Common effects are nose and throat irritation. A severe response includes lung

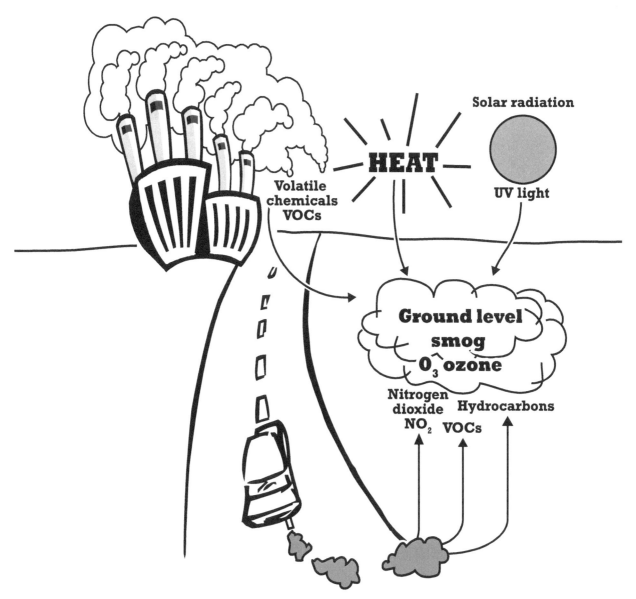

inflammation. If the damage happens over a long period, lung function may be reduced permanently. The elderly, children, and people with lung ailments are particularly vulnerable to smog.

▶ Air Pollution Monitoring System: The Air Quality Index

In cities of the United States, several air pollutants are monitored because of their impact on clean air and human health. Levels of these pollutants are forecast daily. The Air Quality Index (AQI), managed by the Environmental Protection Agency, covers particulate matter, sulfur dioxide, carbon monoxide, nitrogen dioxide, and ozone. Projected pollution concentrations are scaled from 0–500, and the associated threats to human health are also provided. Other countries use similar ratings to inform the public of air pollution levels.

▶ Air Quality Guide for Ozone

Ozone ($\mu g/m^3$) (8-hour average)	Air Quality Index	Cautionary Statement
<116	Good (0–50)	No health impacts are expected when air quality is in this range
117–148	Moderate (51–100)	Unusually sensitive people should consider limiting prolonged outdoor exertion
149–187	Unhealthy for Sensitive Groups (101–150)	The following groups should limit prolonged outdoor exertion: • People with lung disease, such as asthma • Children and older adults • People who are active outdoors
188–226	Unhealthy (151–200)	The following groups should avoid prolonged outdoor exertion: • People with lung disease, such as asthma • Children and older adults • People who are active outdoors • Everyone else should limit prolonged outdoor exertion
227–734	Very Unhealthy (201–300)	The following groups should avoid all outdoor exertion: • People with lung disease, such as asthma • Children and older adults • People who are active outdoors • Everyone else should limit outdoor exertion
>735	Hazardous (>300)	Everyone should avoid all physical activity outdoors

Courtesy of the *Office of Air Quality and Radiation*, EPA, USA www.airnow.gov

▶ What Is Ozone?

Ozone is a colorless gas found in the air we breathe. Ozone can be good or bad depending on where it occurs:

■ Ozone occurs naturally in the earth's upper atmosphere (the stratosphere), where it shields the earth from the sun's UV rays.

■ At ground level, ozone is an air pollutant that can harm human health.

▶ Where Does Ground-Level Ozone Come From?

Ground-level ozone is formed when two types of pollutants react in the presence of sunlight. These pollutants are known as VOCs and oxides of nitrogen. They are found in emissions from the following:

■ Vehicles such as automobiles, trucks, buses, aircraft, and locomotives
■ Construction equipment
■ Lawn and garden equipment
■ Sources that combust fuel, such as large industries and utilities
■ Small industries such as gas stations and print shops
■ Consumer products, including some paints and cleaners

▶ Does My Area Have High Ozone Levels?

Ozone is particularly likely to reach unhealthy levels on hot sunny days in urban environments. It is a major part of urban smog.

Ozone can also be transported long distances by wind. For this reason, even rural areas can experience high ozone levels.

The Airnow Web site at airnow.gov provides daily air quality reports for many areas. These reports use the AQI to tell you how clean or polluted the air is.

▶ How Does Ozone Affect Health?

Ozone can do as follows:
- Make it more difficult to breathe deeply and vigorously.
- Cause shortness of breath and pain when taking a deep breath.
- Cause coughing and sore or scratchy throat.
- Inflame and damage the lung lining.
- Make the lungs more susceptible to infection.
- Aggravate lung diseases such as asthma, emphysema, and chronic bronchitis.
- Increase the frequency of asthma attacks.
- Continue to damage the lungs even when the symptoms have disappeared.
- These effects may lead to increased school absences, visits to doctors and emergency rooms, and hospital admissions.
- Research also indicates that ozone exposure may increase the risk of premature death from heart or lung disease.

▶ Who Is Sensitive to Ozone?

Some people are more sensitive to ozone than others. Sensitive groups include children; people with lung disease, such as asthma, emphysema, or chronic bronchitis; and older adults. Even healthy adults who are active outdoors can experience ozone's harmful effects.

▶ How Can You Keep the Air Cleaner?

- Conserve energy—at home, at work, everywhere. Turn off lights you are not using.
- Car pool or use public transportation. When air quality is healthy, bike or walk instead of driving.
- Combine errands to reduce vehicle trips.
- Limit engine idling.
- When refueling: stop when the pump shuts off. Putting more fuel in is bad for the environment and can damage your vehicle. Avoid spilling fuel. Always tighten your gas cap securely.
- Keep your car, boat, and other engines tuned up.
- Inflate your car's tires to the recommended pressure.
- Use environmentally safe paints and cleaning products whenever possible.
- Follow manufacturers' recommendations to use and properly seal cleaners, paints, and other chemicals so that smog-forming chemicals cannot evaporate.

In this laboratory activity, you will estimate ozone and UV levels as you examine local conditions that contribute to the production of smog. You will use the AQI to determine the health effects associated with the level of ozone that you estimated. During the next week lab, you will explore vehicle emissions directly. You will prepare pure CO_2 from a chemical reaction and use it to create a standard scale to use in quantifying the CO_2 content of collected samples, including exhaled breath and engine exhaust.

REFERENCES

Air Quality Guide for Ozone. Office of Air Quality and Radiation, EPA, USA www.airnow.gov
Air Pollution and Vehicle Emission. *Carolina Biological Supply Company, www.carolina.com*

PROCEDURE

3. Read over the laboratory instructions.
4. Your instructor will distribute one ozone test strip and one UV indicator card to each group.
5. Select a location in which to test the ozone concentration and UV level. Coordinate with other groups so that groups take readings from various indoor and outdoor areas. Keep in mind that ozone test strips are less accurate when placed in direct sunlight or exposed to high wind.
6. Tape the ozone test strip in the chosen location. **Note:** Keep the ozone strip sealed except during the 10 minutes of exposure.
7. Leave the test strip in place for 10 minutes. Continue with ambient temperature and UV testing as you wait.
8. To read the ozone strip, match it to the color scale on the container, which indicates ozone in micrograms per meter cubed ($\mu g/m^3$). Record the location, time, and the ozone level in a row of Data Table.
9. Measure and record the ambient temperature of your testing location.
10. Follow the directions on the UV indicator card. There is a photochemical indicator on the card (inside the sun illustration) that changes when in contact with light in the UV range.
11. Observe the color of the UV indicator and match it to the color scale on the card. Record the time and UV level in your row of Data Table. The card can be used many times. Your instructor may have you take UV estimates on different days to compare various weather conditions and temperatures. The ozone strip works for only one reading.
12. Compare data with the class, filling in the other rows of Data Table.

Lab 7

AIR POLLUTION FROM MOTOR VEHICLE EXHAUST

PRE-LAB READING

Textbook: Chapter 17: Atmospheric Science, Air Quality, and Pollution Control.

ACTIVITIES

1. Preparing Standards
2. Testing CO_2 Concentration in Exhaled Air
3. Measuring Vehicle Emissions

INTRODUCTION

See Introduction for Lab 6.

REFERENCES

Air Pollution and Vehicle Emission. *Carolina Biological Supply Company, www.carolina.com*

ACTIVITY 1 Preparing Standards

MATERIALS

- Bromothymol blue cup
- Vials with cap
- Vinegar cup
- Piece of white paper
- Sodium bicarbonate cup with scoop
- Test tube rack
- Plastic test tube
- Labeling marker and tape
- 10-mL syringes
- Plastic bag
- 1-mL syringe
- Rubber band
- Syringe cap
- Paper towel
- Piece of tubing

PROCEDURE

You will bubble known volumes of carbon dioxide (CO_2) into syringes of bromothymol blue to create color standards to use in assessing the CO_2 volume in unknown samples. When CO_2 is bubbled into solution, it reacts to form carbonic acid, thus acidifying the solution. As the solution becomes more acidic, the bromothymol blue changes from blue (pH 7.6) to green to yellow (pH 6). Each of the standard vials will include 5 mL bromothymol blue with a known volume of pure CO_2 bubbled into it. You will produce the pure CO_2 by mixing vinegar and sodium bicarbonate. The fizz generated by the reaction is CO_2 gas. The balanced equation follows:

$$HC_2H_3O_2(aq) + NaHCO_3(s) \rightarrow NaC_2H_3O_2(aq) + CO_2(g) + H_2O(l)$$

1. Label one 10-mL syringe as *Pure* CO_2.
2. Have one group member hold the syringe while another adds a scoop of sodium bicarbonate to the vinegar in the test tube.
3. About 2 seconds after the reaction starts, hold the syringe so the syringe tip is in the test tube and the syringe is pressed against the rim. No outside air should be let into the tube at this point. Pull the plunger to fill the syringe with 10–12 mL of CO_2. Avoid getting any liquid in the syringe.
4. Seal the Pure CO_2 syringe by screwing on the syringe cap.
5. Add 5 mL bromothymol blue to the remaining 10-mL syringes.

Note: To use the graduations on the syringe, read the volume measurement from the thick black line created by the rubber part of the plunger that is in contact with the solution in the syringe. If your syringe has cc measurements, keep in mind that 1 mL = 1 cc.

6. Blot any excess liquid from the tips of the syringes with a paper towel and set the syringes aside.
7. Your instructor will assign your groups specific volumes of CO_2 to use for preparing the standards. Some groups will prepare odd-numbered volume standards (0.1 mL, 0.3 mL, 0.5 mL, and 0.7 mL), and other groups will prepare even-numbered volume standards (0.2 mL, 0.4 mL, 0.6 mL, and 0.8 mL). Label one vial lid for each of your assigned standards, using labeling tape or stickers and a permanent marking pen.
8. The bromothymol blue reacts with even a small amount of pure CO_2. You will transfer some of the gas to the 1-mL syringe to allow for accurate measurement.
    a. Invert the CO_2 syringe (so that the tip is pointed up) and remove the cap.
    b. Place the piece of tubing on the CO_2 syringe and press the plunger to expel a small amount of air (about 1 mL) to make sure the tubing contains only CO_2.
    c. Attach the free end of the tubing to the fully depressed 1-mL syringe.
    d. Transfer about 1 mL of CO_2 into the 1-mL syringe by pressing the plunger on the CO_2 syringe while gently pulling up on the plunger of the 1-mL syringe.
    e. Disconnect the 10-mL syringe from the tubing and recap. The tubing should still be connected to the 1-mL syringe.
9. Connect the free end of the tubing to one of the bromothymol blue syringes.
10. Hold the syringes so that the tip of the CO_2 syringe is pointing up and the tip of the bromothymol blue syringe is pointing down.

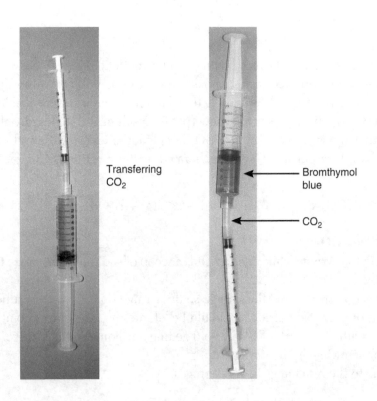

Transferring CO_2

Bromthymol blue

CO_2

11. Slowly depress the CO_2 syringe as you pull up on the plunger of the bromothymol blue syringe. Add 0.7 or 0.8 mL of CO_2, depending on your assigned standards.

12. Disconnect the tubing from the bromothymol blue syringe.

13. Invert the syringe twice and gently expel the bromothymol blue solution into its corresponding vial (labeled 0.7 mL or 0.8 mL). Make sure the vial is completely full before capping it.

14. Repeat steps 9–13 until you have completed the assigned standards. Start with the greater volumes and work to the smaller volumes. You should notice a pattern in color as the volume of CO_2 added to solution decreases. For the smaller volumes, you will not have to refill the 1-mL syringe because the gas can be used for preparing multiple standards.

15. Follow your TA instructions to compare and organize the standards prepared by the class. Observe how your standards compare with those of other groups. Some of the standard vials may be discounted if their colors deviate from the obvious color gradient. You will use these approved standards for comparison with samples that you collect from various exhaust sources.

Lab 7

ACTIVITY 2 Testing CO_2 Concentration in Exhaled Air

PROCEDURE

1. You should have one 10-mL syringe left with 5 mL of bromothymol blue. If not, prepare another one.
2. Collect exhaled air in a plastic bag.
   a. Bunch the open end of the bag between your thumb and forefinger.
   b. Spread the opening of the bag so that you can blow air into it.
   c. Have another group member squeeze any existing air out of the bag.
   d. Exhale a couple of breaths of air between your thumb and forefinger into the bag. Clamp the bag shut between breaths.
   e. Secure the exhaled air inside the bag by tying it off with a rubber band.
3. Label one vial with your name and "breath" on the cap.
4. Collect a known volume of exhaled air.
   a. Carefully unwrap the bag.
   b. Hold an empty, depressed 10-mL syringe inside the bag, careful not to let any air escape.
   c. Pull the plunger to collect about 10 mL of exhaled air in the syringe.
   d. Close the bag and secure it with a rubber band.
   e. Place the syringe cap on the syringe.
5. Use the technique that you used with the standards to bubble the breath sample into the bromothymol blue. (Because larger volume of exhaled breath is needed to produce a pH change equivalent to that resulting from pure CO_2, you do not need to transfer the CO_2 to the smaller syringe.)
   a. Place the piece of tubing on the sample syringe and expel a small amount of air (about 1 mL) to make sure the tubing contains only the sample gas.
   b. Connect the free end of the tubing to one of the bromothymol blue syringes.
   c. Hold the apparatus so that the syringes are perpendicular to the ground, with the tip of the sample syringe pointing up and the tip of the bromothymol blue syringe pointing down.
   d. Slowly depress the sample syringe as you pull up on the plunger of the bromothymol blue syringe, adding 1 mL of gas. Mix the contents by inverting the entire system twice.
   e. Continue adding 1 mL at a time and mixing until there is a clear color change. Do not add more than 7 mL.
   f. Record in Data Table 7.2 the volume of sample bubbled into solution.
   g. Transfer the solution in the syringe to the appropriately labeled vial and cap the vial.
6. Collect your standards and team with another group who was assigned the opposite set (if available). You should have eight standards (minus any error vials that were discarded).
7. Compare the exhaled breath samples with the standards. Record your observations in Data Table 7.1.

8. Calculate the percentage of CO_2 in each gas sample.
   Example calculation: If your sample matches the 0.1-mL CO_2 standard, then it has about 0.1 mL of CO_2. To calculate the percentage of CO_2 in the sample, divide 0.1 by the total volume of sample gas bubbled into bromothymol blue. If the volume of sample gas was 2 mL, use the following calculation:

$$0.1 \text{ mL } CO_2/2 \text{ mL} = 0.05 \times 100 = 5\% \ CO_2 \text{ in the exhaust sample}$$

9. Enter the estimated CO_2 concentration into Data Table 7.1.

Lab 7

ACTIVITY 2 Testing CO_2 Concentration in Exhaled Air

Student Name: _____   Lab Group: _____

TA: _____   Lab Date/Section: _____

DATA

TABLE 7.1. CO_2 Level in Exhaled Gas

Sample Source	Volume of Sample Added to 5 mL Bromothymol Blue	Matched Pure CO_2 Sample	Estimate of CO_2 Concentration

ACTIVITY 3 Measuring Vehicle Emissions

MATERIALS

- Standards prepared in Activity 1
- 2 syringe caps
- 2 plastic bags
- Piece of tubing
- 2 folders to use as funnels
- Vials with caps
- 3 rubber bands
- Labeling marker
- 6 10-mL syringes
- Heat-resistant gloves

▶ Procedure for Collecting Exhaust

1. Open the folder and begin rolling from a corner to get a long funnel. (A longer funnel keeps your hands farther away from hot exhaust pipes.) Secure a funnel with a rubber band.
2. Prepare your bag for collecting exhaust:
   a. Label the bag with your group name and sample source.
   b. Loop another rubber band around the narrow end of the funnel.
   c. Wrap the opening of the bag tightly around the narrow end of the funnel and secure it with one of the rubber bands.
   d. After examining the exhaust port that you are going to sample, adjust the shape of the funnel so that the wider end can fit over the exhaust without any gas escaping.
3. Collect the exhaust:
   a. Start the engine.
   b. **Use caution. The tailpipe is very hot. Heat-resistant gloves are recommended.**
      Hold the funnel by the narrow end so that your hand is far from the exhaust pipe itself. Place the wider end against the mouth of the tail pipe, and let the bag fill with exhaust (2–3 seconds).
   c. Secure the bag of exhaust by folding over the top and sealing it with the rubber band that had attached it to the funnel.
   d. Repeat this process with your second exhaust source.
4. Return to your laboratory to analyze the exhaust.

▶ Procedure for Comparing CO₂ Concentrations in Exhaust

1. Prepare four 10-mL syringes, each with 5 mL bromothymol blue.
2. Label the four syringes with the sample name and "a" or "b." Each sample will be tested in duplicate.
3. Label the four vials with the sample name and "a" or "b" on the lid.

4. Collect a known volume of sample 1.
   a. Carefully unwrap the bag for sample 1.
   b. Hold an empty, depressed 10-mL syringe inside the bag, careful not to let any exhaust gas escape.
   c. Pull the plunger to collect about 10 mL exhaust in the syringe.
   d. Close the bag and secure it with a rubber band.
   e. Place the cap on the syringe.
5. Use the basic technique that you used with the standards to bubble the sample into the bromothymol blue. Because the exhaust is not pure CO_2, a larger volume will be needed to result in a similar pH change insolution.
   a. Invert the CO_2 syringe (so that the tip is pointed up) and remove the cap.
   b. Place the piece of tubing on the sample syringe and expel a small amount of air (about 1 mL) to make sure the tubing contains only exhaust gas.
   c. Connect the free end of the tubing to one of the bromothymol blue syringes.
   d. Hold the syringes so that they are both perpendicular to the ground. The tip of the sample syringe is pointing up and the tip of the bromothymol blue syringe is pointing down.
6. Bubble sample 1 through one of the syringes of bromothymol blue. This is sample 1a.
   a. Slowly depress the sample syringe plunger as you pull up on the plunger of the bromothymol blue syringe.
   b. Slowly add 1 mL of sample to the bromothymol blue and invert the entire system twice.
   c. Continue adding 1 mL at a time and inverting twice until you see a clear color change. Do not add more than 7 mL.
   d. Record the volume of sample bubbled into solution in your Data Table.
   e. Transfer the solution in the syringe to the appropriately labeled vial and cap the vial.
7. Repeat step 6 by bubbling the same volume of sample 1 gas into a new bromothymol blue syringe as a duplicate test. This is sample 1b.
8. Repeat steps 6 and 7 for samples 2a and 2b.
9. Make sure all the sample vials are clearly labeled. Compare your samples to the set of class standards to estimate the CO_2 concentrations of your samples. Record your observations in Data Table 7.2.

Lab 7

ACTIVITY 3 Measuring Vehicle Emissions

Student Name: _____   Lab Group: _____

TA: _____   Lab Date/Section: _____

DATA

TABLE 7.2. Comparative Engine Emissions

Make, Model, Year of Vehicle	Number of Cylinders	Type of Fuel		Volume of Sample Added to 5mL Bromothymol Blue	Matched Pure CO_2 Sample	Estimate of CO_2 Concentration
			a			
			b			
			a			
			b			

Lab 7

Air Pollution from Motor Vehicle Exhaust

Student Name: _____ Lab Group: _____

TA: _____ Lab Date/Section: _____

DISCUSSION AND CONCLUSION

1. Determine factors that may influence the volume and composition of vehicle exhaust and describe how they do so.

   _____

   _____

   _____

   _____

2. What were your expectations or predictions before you compared the measurements of the emission from two vehicles? How did the prediction change or not change?

   _____

   _____

   _____

   _____

3. What conclusion can you draw from the results of this activity?

   _____

   _____

   _____

   _____

4. Do you think choosing public transportation when possible would have an effect on air pollution in our area? What prevents you from taking a bus or a bike instead of driving a car?

   _____

   _____

   _____

   _____

Lab 8

GLOBAL CLIMATE CHANGE

PRE-LAB READING

Textbook: Chapter 18: Global Climate Change.

ACTIVITY

Investigating Global Climate Change

Lab 8

ACTIVITY 1 Investigating Global Climate Change

OBJECTIVES

After completing this experiment, you should be able to:

- Understand the term Global Climate Change
- Understand how human activity has affected our global climate
- Analyze and interpret scientific data
- Discuss the concept of Global Climate Change

INTRODUCTION

What is Global Climate Change? We often hear it mentioned, but do we really know why it happens, what the consequences are and what caused it? Most importantly though, is it real? In order to understand Global Climate Change, we first need to have some basic knowledge about the atmosphere.

The sun makes life on Earth possible. The sun emits solar radiation (high energy such as visible and UV light) of which ~1/3 is reflected back into space by clouds, the atmosphere and the surface of the planet. The remaining radiation is absorbed by clouds and the surface of the planet. These both become warmer and start to emit low-energy infrared radiation back toward the atmosphere. Infrared radiation does not pass through the atmosphere as easily as visible or UV radiation. It is absorbed by gases in the atmosphere, which consequently warm. When these gases warm, they emit infrared radiation. Some of this radiation goes out into space while the rest is emitted back toward the surface of the Earth. **The Greenhouse Effect** is the absorption of infrared radiation by gases in the atmosphere and the re-radiation of energy back towards Earth. This re-radiation results in the planet becoming warmer [3, 7].

The gases in the atmosphere that absorb infrared radiation are called **greenhouse gases**. Gases such as water vapor, carbon dioxide, methane, nitrous oxide and chlorofluorocarbons (CFCs) are all greenhouse gases. Sources of these gases are both natural and anthropogenic. **Natural sources** include: (1) Volcanic eruptions that release CO_2; (2) Decomposition of organic matter occurring under low oxygen conditions (e.g., wetlands, bacterial breakdown in the gut of animals) releasing Methane; (3) Low oxygen conditions in the bottom of wetlands, lakes and oceans, which produce nitrous oxide, a process known as denitrification; and (4) Evaporation of liquid water from land and water bodies as well as evapotranspiration in plants. Manmade or **Anthropogenic sources** include: (1) Burning of fossil fuels, causing release of carbon dioxide; (2) Agriculture, which causes a large release of methane from manure and digestion of plant matter in cattle. Agriculture also contributes to the release of nitrous oxide and methane in over-irrigated pastures (e.g., rice fields that are low in oxygen) and from denitrification in fertilized soil; (3) Deforestation, which removes the trees that sequester a large amount of carbon dioxide from the atmosphere; (4) Landfills, which create low oxygen conditions and hence release methane; and (5) Industry, which releases new and manmade greenhouse gases such as chlorofluorocarbons (CFCs) found in refrigerators and air conditioners [7].

Of all greenhouse gases, carbon dioxide remains the greatest contributor to the greenhouse effect due to both its concentration and its duration in the atmosphere [7]. In the late 1700s, the CO_2

concentration in the atmosphere was around 280 ppm, but by 2002, CO_2 concentrations had risen to about 373 ppm. Thus, over a time period of 300 years, the concentration of CO_2 in the atmosphere has increased by ~33%. Figure 7.1 shows the atmospheric CO_2 levels in the atmosphere in the last 50 years (**Fig. 8.1**). As you can see, the levels are still dramatically increasing. This increase in CO_2 along with other greenhouse gases has caused the annual average U.S. temperature to increase by 0.6°C during the twentieth century [1]. It is important to note that controversy still exists as to the origin of this increase (human activities versus naturally occurring). To resolve this controversy, scientists have determined CO_2 levels in the atmosphere dating back 400,000 years (**Fig. 8.2**). As no records are available from that time, scientists are determining

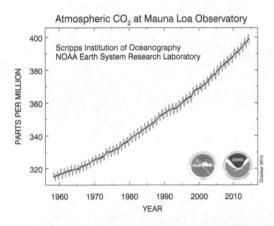

FIGURE 8.1.
Average and monthly changes in atmospheric carbon dioxide over time. Solid line denotes overall measurements while grey oscillating line denotes monthly averages. Courtesy of NOAA. Question: Why does CO_2 concentration oscillate over time?

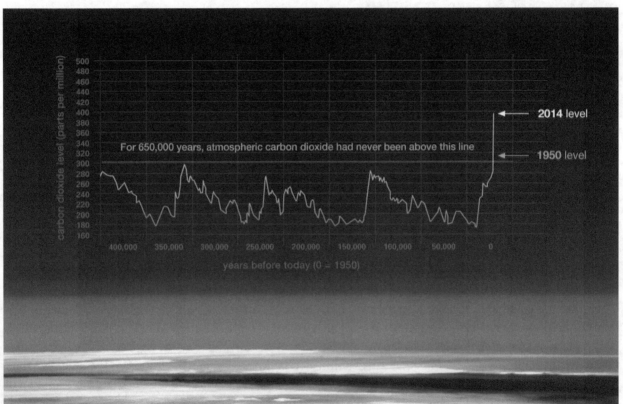

FIGURE 8.2.
Atmospheric carbon dioxide concentration over time. Comparison of atmospheric samples contained in ice cores and more recent direct measurements provide evidence for an increase of atmospheric carbon dioxide since the Industrial Revolution.

atmospheric gases trapped in ancient ice by collecting ice cores. One ice core can cover a time span of up to 500,000 years! By looking at ice cores, scientists have determined that CO_2 emissions dramatically increased following the Industrial Revolution (~1800s) when the burning of coal and oil exponentially increased [7].

The United Nations Intergovernment Panel on Climate Change (IPCC) has projected a 1.4–5.8°C (2.5–10.4°F) increase in global temperature by the year 2100. It may not sound like a large increase, but how will this warming affect our environment and all living organisms? Environmental changes are predicted to include melting of the polar ice caps, melting of glaciers and permafrost, rising sea levels, increased heat waves and cold spells, change in precipitation patterns (increased rainfall and drought) and change in ocean currents, as well as increased storm intensity [7]. Living organisms (plants and animals, including humans) will need to cope with increased temperatures and a shift of seasons. This will affect agriculture as traditional crops and livestock may not be able to survive in warmer climates. On the positive side, agriculture may be expanded to areas currently too cold to grow crops and keep livestock. The disadvantage of increasing temperatures is the enhanced risk of animal and plant pests such as weed and insect pests. These may flourish and move into areas where they are currently not present. For humans, disease-carrying organisms, such as mosquitoes, may also thrive and expand their range and spread malaria, dengue fever and yellow fever [3, 5]. Warmer temperatures may mean that some wild plants and animals are able to expand their habitat ranges and migrate to more hospitable areas. However, survival of most animal species depends on whether their food supply is able to adapt to changes in climate. For instance, some species may not cope with climate change because their response differs from the response of organisms at lower levels of the food chain [6].

Below is a list predicted by scientists about the effects of climate change on the US [1, 7]:

- Mountaintop ecosystems will become more endangered.
- Southeastern U.S. forests will break up into savanna/grassland/forest mosaics.
- Northeastern U.S. forests will lose native sugar maples.
- Loss of coastal wetlands and real estate due to sea level rise will continue.
- Snow-pack reduction will continue.
- Water shortages will worsen.
- Melting permafrost will undermine Alaskan buildings and roads. Melting permafrost will also create low-oxygen conditions in wet soils causing a release of methane when thawed organic matter in the tundra starts breaking down (positive feedback).
- Skiing and other winter sports will decline.
- Air conditioning use will increase.
- Some crop and forest productivity may increase.
- Average U.S. temperatures will increase 3–5°C in the next 100 years.
- Some ecosystems, including alpine meadows and barrier islands, may disappear altogether while others, such as Southeastern U.S. forests, will change dramatically.
- Drought, floods and snow-pack changes will give rise to environmental and social problems.
- Impacts on U.S. food supply are hard to predict. Drought and other factors may present problems while longer growing seasons may benefit some crop producers.
- Forest growth may increase in the short term, but in the long term, drought, pests and fire may also alter forest ecosystems.
- Greater temperature extremes will cause increased health problems and mortality among human populations. Some tropical diseases will also spread north into temperate latitudes.

In this lab, you will be presented with a simulated scenario involving actual scientific data. You will be the scientist who has set out to determine whether the Earth's climate is indeed changing. Your scientific team has already been in the field, collecting the data you need to support or reject your null hypotheses. It is now your job to plot, interpret and discuss the data collected from the field.

To prepare for this lab, you can check out the following web sites that have questions and answers about climate change. All of these web sites have updated information about causes and effects of Global Climate Change, as well as information about scientific research currently being conducted.

http://www.usgs.gov/global_change/
http://www.epa.gov/climatechange/index.html
http://www.ncdc.noaa.gov/oa/climate/climateextremes.html
http://climate.nasa.gov

MATERIALS

Pencil (not pen)
Graph paper
Ruler
Data tables with data from *http://climate.nasa.gov*

METHODS

You will be provided with handouts of actual data (numbers) collected by scientists. The parameters measured are all key indicators used by scientists to determine whether the Earth's climate is changing. The parameters the scientists measured are:

1. Atmospheric CO_2 Levels.
2. Global Surface Temperature.
3. Arctic Sea Ice Cover.
4. Sea Level Change.

Each parameter was determined over time, with the level of CO_2 dating back the furthest (~400,000 years). Ancient atmospheric CO_2 levels were indirectly determined by studying gases trapped in ancient ice. Long tubes of ice were obtained by drilling deep into the ice and extracting ice cores. From 1950 and onwards, CO_2 levels were determined directly. Global surface temperature was determined as the change in global surface temperature relative to average temperatures from 1951 to 1980. Hence, you will be plotting Temperature Anomaly instead of actual temperatures. The Arctic Sea ice cover was determined by satellite observations and this method was also used to estimate changes in sea level. The data you will be provided were obtained from graphs on the web site *http://climate.nasa.gov* and rounded to the nearest whole number to help you plot the graphs.

▶ **Research hypotheses (H₁)**

Before your team went into the field, you developed the following research hypotheses:

H_{1A}:   **Atmospheric CO_2 levels have increased over time.**
H_{1B}:   **Global surface temperature has increased over time.**

H_{1C}:   **Atmospheric CO_2 levels are responsible for the rise in global temperature.**

H_{1D}:   **Arctic Sea ice cover has decreased over time.**

H_{1E}:   **Sea levels have increased over time.**

Your null hypotheses are the following:

H_{0A}:   **Atmospheric CO_2 levels have not increased over time.**

H_{0B}:   **Global surface temperature has not increased over time.**

H_{0C}:   **Atmospheric CO_2 levels are not responsible for the rise in global temperature.**

H_{0D}:   **Arctic Sea ice cover has not decreased over time.**

H_{0E}:   **Sea levels have not increased over time.**

In order to be able to support or reject the null hypotheses, you need to analyze the data you collected in the field.

DATA ANALYSIS

A. Table 1: Plot atmospheric CO_2 levels in part per million (ppm) (Y-axis) as a function of time (X-axis). *Hint:* You will have to create a break \\ in your X-axis as you have a gap of 50,000 years in your data table.

B. Table 2: On a separate graph, plot temperature anomaly in degrees Celsius (°C) (Y-axis) as a function of time (X-axis). For both graphs, connect your data points with a line of best fit (do not use a ruler, draw the line by free hand).

C. Table 3: Plot temperature (Y-axis) as a function of CO_2 levels (X-axis). After you have plotted the graph, connect the dots by free hand.

    One of the questions you wanted to answer was whether atmospheric CO_2 levels are responsible (the cause) for rising surface temperatures. To answer this, you want to determine whether there is a correlation between surface temperatures and CO_2 levels in the atmosphere. Use a ruler and draw a trend line across all your points on the graph. Your instructor can help you to determine the best trend line.

D. Table 4: Plot the area of Arctic Sea Ice Cover (in millions of square kilometers) (Y-axis) as a function of time (X-axis).

E. Table 5: Plot the change in sea levels in millimeters (mm) on the Y-axis as a function of time (X-axis).

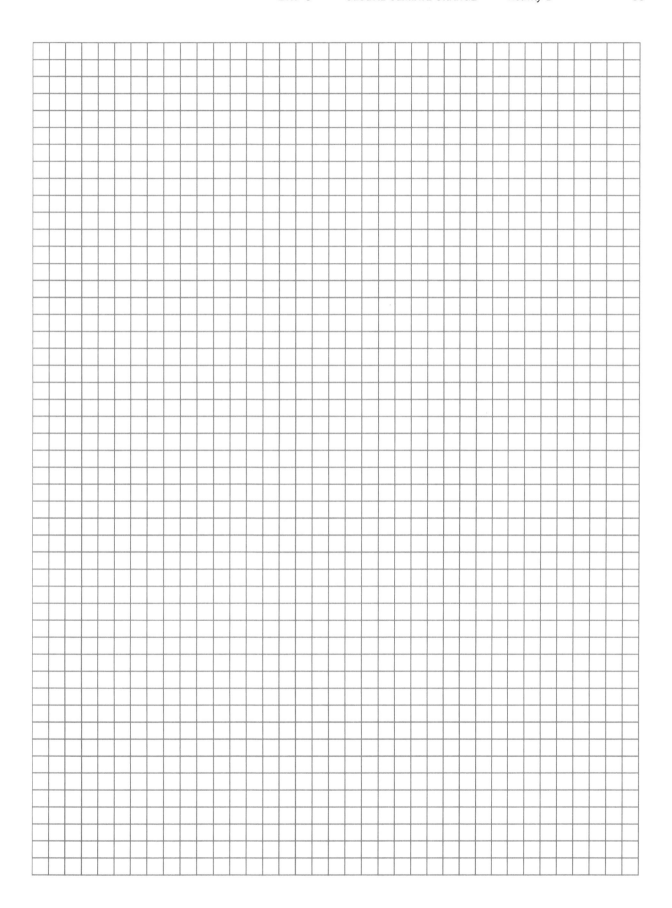

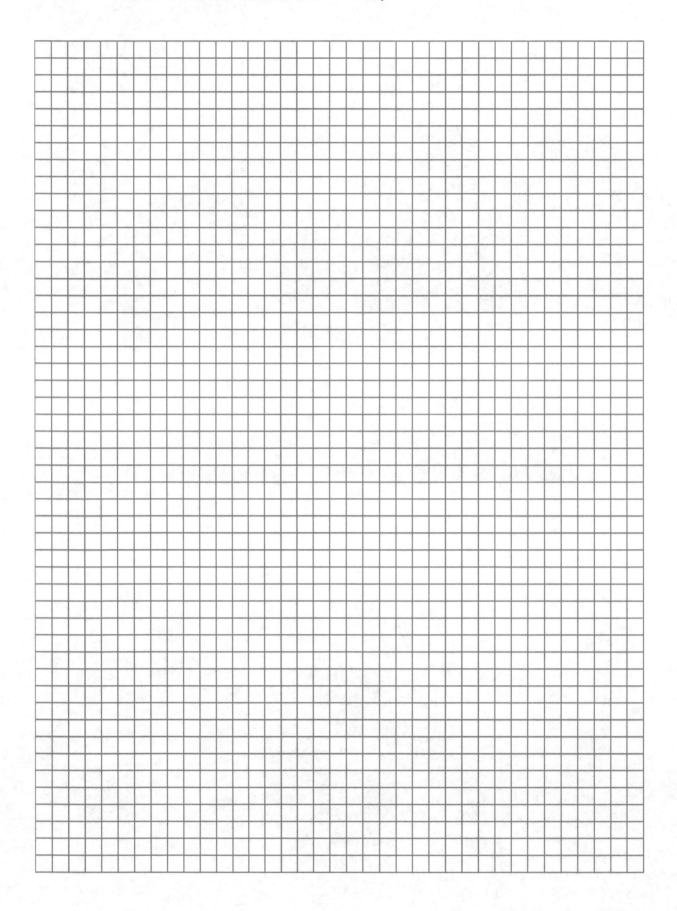

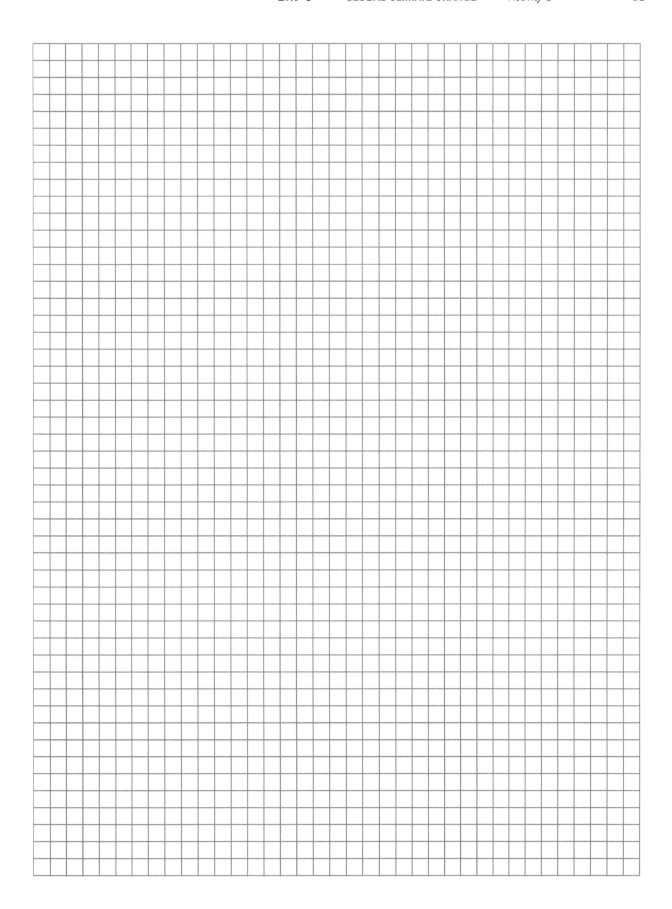

REFERENCES

Scott, B., and J. Withgott. 2005. *Environment: The Science Behind the Stories.* Pearson Education, Inc., publishing as Benjamin Cummings.

Kunzig, R. 2011. "Seven Billion." Special Series. *National Geographic Magazine,* January, 42–63.

Berg, L. R., and M. C. Hager. 2007. *Visualizing Environmental Science.* Hoboken. NJ. USA: Wiley and Sons.

Cannon, R. J. C. 1998. "The implications of predicted climate change for insect pests in the UK, with emphasis on non-indigenous species." *Glob. Chang. Biol.* 4: 785–796.

Siebert, C. 2011. "Food Ark. Seven Billion." Special Series. *National Geographic Magazine,* July, 108–132.

Both, C., S. Bouwhuis, C. M. Lessells, and M. E. Visser. 2006. "Climate Change and Population Declines in a Long-distance Migratory Bird." *Nature.* Vol. 441.

Friedland, A., R. Relyea, and D. Courard-Hauri. 2012. "Environmental Science: Foundations and Applications." Chapter 19. In *Global Change: Climate alteration and global warming.* New York: W. H. Freeman.

Lab 8

ACTIVITY 1
Global Climate Change

Student Name: _____ Lab Group: _____

TA: _____ Lab Date/Section: _____

▶ Interpretation of Results

Once you have plotted your results, look at the graphs you produced. Analyze and briefly explain why your data either supported or rejected your null hypotheses and therefore proved your research hypotheses true or false:

H_{1A}: _____

_____

_____

_____

H_{1B}: _____

_____

_____

_____

H_{1C}: _____

_____

_____

_____

H_{1D}: _____

_____

_____

_____

H_{1E}: _____

_____

_____

_____

Lab 9
GLOBAL WARMING

PRE-LAB READING

Textbook: Chapter 18: Global Climate Change.

ACTIVITIES

1. Biodegradation Project Completion
2. Movie: "An Inconvenient Truth"

Biodegradation Project Completion

PROCEDURE

See instruction in Appendix I: Activity 3.

ACTIVITY 2 Global Warming—"An Inconvenient Truth"

OBJECTIVES

- ■ Understand the concept of the greenhouse effect.
- ■ Understand the causes of global warming.
- ■ Understand the potential consequences of global warming.
- ■ Understand what actions could be taken to mitigate the global warming.

HYPOTHESIS

- ■ None.

MATERIALS

- ■ "An Inconvenient Truth" DVD

PROCEDURE

1. The class will view the movie "An Inconvenient Truth."
2. Prior to viewing the movie, review the information you are for on the data sheet so that you will know what to be listening for during the movie.
3. Answer the questions that you can during the movie.
4. After the movie, discuss its strengths and weaknesses with your group.
5. Record your response in Tables 9.13, 9.14, and 9.15 on the data sheet.

From *Environmental Science Lab Manual and Notebook, Volume 2: The Issues* by Kim Bowling Largen. Copyright © 2010 by Kendall Hunt Publishing Company. Reprinted by permission.

Lab 9

ACTIVITY 2

Global Warming—"An Inconvenient Truth"

Student Name: _____   Lab Group: _____

TA: _____   Lab Date/Section: _____

DATA

TABLE 9.1. List Five Words/Concepts That Are Important to Understanding Global Warming and
Definitions/Explanations for Each

	Word/Concept	Definition/Explanation
a.		
b.		
c.		
d.		
e.		

Lab 9

ACTIVITY 2 — Global Warming—"An Inconvenient Truth"

Student Name: _____ Lab Group: _____

TA: _____ Lab Date/Section: _____

DATA

TABLE 9.12. Four *Weaknesses* of the Film's Message about Global Warming, and Supporting Reasons, Identified by Group Members During Post-Film Group-Level Discussion

	Weakness	Reason
a.		
b.		
c.		
d.		

Lab 9

ACTIVITY 2

Global Warming—"An Inconvenient Truth"

Student Name: _____   Lab Group: _____

TA: _____   Lab Date/Section: _____

DATA

TABLE 9.13. Four *Strengths* of the Film's Message about Global Warming, and Supporting Reasons, Identified by Group Members During Post-Film Group-Level Discussion

	Strength	Reason
a.		
b.		
c.		
d.		

Lab 9

ACTIVITY
2

Global Warming—"An Inconvenient Truth"

Student Name: _____   Lab Group: _____

TA: _____   Lab Date/Section: _____

DISCUSSION & CONCLUSIONS

For full credit, questions should be answered thoroughly, in complete sentences, and legibly.

1. *Did* the movie convince you that global warming is occurring as a result of human activities? Explain *why* or *why not?*

_____

_____

_____

_____

2. If the movie convinced you that global warming is the result of human activities, *did* it motivate you to take personal action? *Why* or *why not?*

_____

_____

_____

_____

3. *What* personal actions do you think you would/could realistically take?

_____

_____

_____

_____

4. *Would* you purchase a hybrid or electric car? *Why* or *why not?*

_____

_____

_____

_____

5. *Are* there energy options for the average citizen that do not contribute to global warming?

_____

_____

_____

_____

6. *Is* global warming an urgent issue in need of remedial action? *Why* or *why not?*

_____

_____

_____

_____

Lab 10

ENERGY SOURCES.
PART 1

PRE-LAB READING

Textbook: Chapter 19: Fossil Fuels; Chapter 20: Conventional Energy Alternatives; Chapter 21: New Renewable Energy Alternatives.

OBJECTIVES

- Learn about renewable and nonrenewable energy resources.
- Examine different energy technologies, and analyze pros and cons of each.
- Test the properties of the gas released when coal shavings are heated.
- Set up a working solar panel.
- Build a wind turbine and discuss ways to improve its design.
- Explore hydrogen fuel cells by splitting water molecules and testing the properties of gases produced during the chemical process.

ACTIVITIES

1. Coal Gas
2. Solar
3. Wind
4. Battery

INTRODUCTION

▶ Background

Virtually, every moment of every day, even when you are asleep, you are in some way or other consuming energy.

Some of this consumption is obvious and intentional, but often you may not even be aware of it. Have you ever thought about where the energy comes from to power not only vehicles and light bulbs, but also cell phones, computers, refrigerators, televisions, and game consoles? Residents of the United States use a phenomenal amount of energy every day, and that amount continues to rise. As the earth's population increases and as nations become more technologically advanced, the demand for energy continues to grow.

Over the last 200 years, nonrenewable energy sources such as coal and oil have provided most of the world's energy. During this time, these resources have become more expensive to extract. One inevitable problem with nonrenewable fuel sources is that they will eventually run out. Not only are these resources limited, but their use generates many pollutants, including greenhouse gases linked to global warming.

Accessibility and quality of resource deposits are important factors in coal and oil extraction operations. As resources are exhausted, these industries are increasingly forced to consider locations that are less ideal and deposits that are of poorer quality. As challenges like these mount, increased emphasis is being placed on the value of energy generated from sources that are sustainable over the long term—energy sources such as the sun, the wind, moving water, and biofuels. Not only are these energy sources renewable, they typically produce less pollution than the burning of fossil fuels for the same purpose. However, each of these technologies is accompanied by a specific set of challenges and concerns.

▶ The Future of Energy

As people realize that fossil fuels are not going to last forever, that energy demand will continue to increase, and that existing technologies cause excessive environmental damage and pollution, many new technologies are being developed with the goal of addressing these problems. New energy sources are being explored, and new technologies are being applied to improve the safety, efficiency, and cleanliness of existing energy sources. It is estimated that the world population will surpass 9 billion by the year 2050.

No single technology can effectively supply enough energy for the world, and every source has advantages and disadvantages. The likely scenario for the future is that a multitude of new energy technologies will be used simultaneously and in conjunction with one another.

Over the course of this and the next labs, you will explore and experiment with a variety of nonrenewable and renewable energy sources, specifically coal gas, solar, wind, battery, natural gas, geothermal, water, and hydrogen power.

Lab 10

ACTIVITY 1 Coal Gas

The United States has a tremendous supply of coal. Some forecast the supply of coal will last another 100 to 200 years. Coal has been an important energy source for the United States, first for heating, transportation, and manufacturing, and later for large-scale generation of electricity. Burning coal releases carbon dioxide, as well as nitrogen and sulfur gases. Carbon dioxide is a major greenhouse gas that contributes to climate change. Nitrogen and sulfur gases mix with water vapor in the air and produce weak acids. Over widespread areas in the eastern U.S., acid rain damages forests and pollutes streams and lakes, killing many aquatic organisms. The acid also deteriorates limestone, a component in many buildings and bridges.

One alternative to burning coal is to gasify it. Heating coal and injecting steam into it causes the coal to break down into its component molecules, including carbon monoxide and hydrogen gas. Both of these gases can be used as fuel. The remaining components, which normally cause most of the pollution associated with coal, are captured during gasification and can be sold for other uses. Coal gasification helps to reduce the volume of pollutants produced.

MATERIALS

- Small piece of coal
- Test tube
- Metal file
- Paper towel
- Support stand with buret clamp
- Bunsen burner
- Several toothpicks
- 1-hole stopper with pipet inserted
- Matches
- Heat-resistant gloves
- Safety goggles

PROCEDURE

1. Set up the support stand with a buret clamp above a Bunsen burner.
2. Using the metal file, shave off a small pile of coal onto a paper towel. There should be just enough to cover the bottom of a test tube.
3. Carefully dump the coal shavings into a test tube.

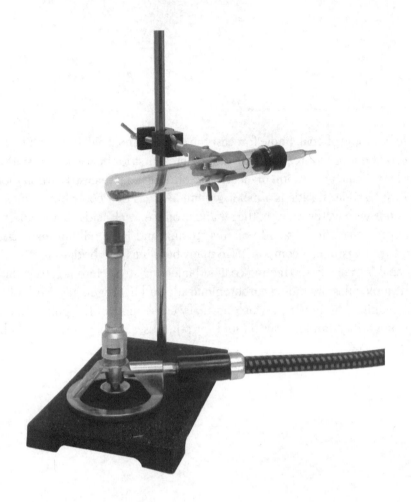

4. Cap the test tube with the stopper and place it in the buret clamp. The tube should be positioned horizontally, with the coal shavings directly over the burner.

5. While wearing safety goggles and heat-resistant gloves, light the burner and, if necessary, lower the test tube until the flame is touching the glass at the coal shavings.

6. While the gas fills the test tube and begins to exit through the end of the pipet, light a toothpick. Hold the lit toothpick just below the tip of the pipet until it ignites the escaping gas. This may take several attempts.

7. Observe what happens not only to the gas coming out of the test tube, but also to the gas inside the tube. Be sure to completely extinguish the toothpick and turn off the burner.

8. Allow the test tube to cool completely before removing it from the buret clamp. Once cooled, dispose of the test tube and coal shavings as directed by your teacher.

9. Answer the questions for the *Coal Gas* on Discussion and Conclusion page.

Lab 10

ACTIVITY 2   Solar

Most of the sun's energy that hits the earth is reflected back into space. Photosynthesizers capture less than 1% of the solar energy that reaches the earth, thereby establishing the basis of the planet's food web.

Solar energy is abundant and clean, and its use has gained momentum in recent decades. There are many ways to harness energy from the sun. On some homes, plastic water pipes are run along the roof and covered with black tarps to absorb sunlight during the day and heat water. This reduces reliance on a water heater. In some parts of the world, people utilize solar cookers, which have mirrors to concentrate the sun's rays onto a small area, producing heat for cooking food.

On a larger scale, solar powered electricity plants are being built all over the world. Some plants use a field of mirrors to direct the sun's rays to a central boiler to produce steam, which rotates turbines and generates electricity. Photovoltaic (PV) cells are used in solar power generating plants as well as in homes. These cells are designed to capture the sun's energy and convert it directly into electricity. These cells are widely used to power many devices, including calculators, clocks, and landscaping lights.

Currently, one problem with solar power is the cost of building the PV cells. Some of the materials, such as silicon, are in high demand, causing the cells to be expensive. Exacerbating this problem is the fact that a single solar cell does not produce much energy; numerous cells must be installed in a circuit to produce enough electricity to be useful. New varieties of PV cells are being developed to overcome these obstacles. While solar energy holds a great deal of potential in locations that receive a lot of strong sunlight, its viability in locations having less ideal conditions (i.e., in extreme latitudes or constant shade) diminishes accordingly.

MATERIALS

- Solar cell
- Compass
- 2' wire
- Tape
- 2 alligator clips (black and red)
- Light source (lamp or LED)

PROCEDURE

1. Assemble a galvanometer, a device that can measure the relative strength of an electric current. When the galvanometer is aligned with the earth's magnetic field and subjected to a perpendicular current from another source (creating a perpendicular magnetic field), the angle of deflection of the compass needle from North gives an indication of the strength of the unknown current. The greater the deflection, the stronger the magnetic field and electric current producing it.

   a. Orient the strand of wire in a North–South direction across the face of the compass, with one end of the wire extending slightly from the edge of the compass marked "N" (North). Wrap the wire entirely around the compass from top to bottom (i.e., from North to South) a total of five times. One end of the wire should project North, while the other end projects South.

   b. Tape the wire securely to the compass.

   c. Place the compass on the lab bench, and rotate it until the compass needle points to North. Then, tape the compass to the lab bench.

2. Attach the black alligator clip to the end of the wire at the North end of the compass. Attach the red alligator clip to the end of the wire at the South end of the compass.

3. Attach the free end of the red alligator clip to the red wire on the PV cell. Attach the free end of the black alligator clip to the black wire on the PV cell.

4. Turn on the light source and shine light onto the PV cell.
5. Observe what happens to the compass needle. Measure in degrees how far the needle deflects from North.
6. Change the angle of the light shining on the PV cell. Observe how the angle of the compass needle changes.
7. Alter the distance between the light source and the PV cell. Observe how the angle of the compass needle changes.
8. Disassemble the solar cell circuit and galvanometer.
9. Answer the questions for the *Solar* activity on Discussion and Conclusion page.

ACTIVITY
3
Wind

For thousands of years, man has harnessed the power of the wind. Since ancient times, sailing vessels have used the wind for propulsion, and the mechanical energy of windmills has been used to grind grain and to pump water for centuries. The addition of a generator enables a windmill to produce electricity through a process called electromagnetic induction. Like sunlight, wind is a clean, inexhaustible, and abundant source of energy. After the initial cost of building and erecting turbines and transmission lines, wind energy is relatively cost-effective.

Some of the controversial issues surrounding wind power are noise and light pollution and the threat it presents to wildlife. Wind turbines are often located in rural areas, where residents unaccustomed to the background noise have reported interrupted sleep and anxiety attributed to the presence of the turbines.

In some cases, light pollution is also an unwelcome side effect. In the United States, any structure built taller than

200 feet must be equipped with warning lights as an aviation safety precaution. On a large wind turbine farm, this requirement can result in hundreds or thousands of incessantly blinking red and white lights.

Furthermore, maintaining and repairing these towering structures can present its own set of logistical challenges—especially for offshore and mountaintop wind farm locations.

Conservationists are concerned with the effects of wind turbines on wildlife populations. The migratory routes of many birds and tree bats follow coastal mountain ridges characterized by sustained high winds— the same areas that are best suited for wind farms. There have been reports of significant numbers of birds and bats killed by the towers and turbines. Wildlife biologists continue to monitor the extent of this problem. In some areas, the mortality rate is deemed troubling, but perhaps less significant in comparison with the number of migratory birds killed by power lines or motor vehicles. Along with the noise issue, wildlife protection has become one of the issues most taken into account in the location and design of new wind farms.

MATERIALS

- 12-hole crimping hub
- 10 wooden dowel rods
- Support stand with buret clamp
- DC motor
- Fan
- 2 alligator clips (black and red)
- Compass
- 2' wire

- Tape
- Several sheets of paper
- Scissors

PROCEDURE

1. Set up a support stand with a buret clamp, as shown. Be sure to place the clamp high on the stand so that when attached, the turbine blades will clear the base of the stand.
2. Attach the hub to the small DC motor by inserting the shaft of the DC motor into the hole in the center of the crimping hub. Unscrew the front part of the hub.
3. Discuss among your group how many dowel rods to attach to the hub, that is, how many blades should be there on the turbine.
4. Insert the dowel rods into the hub. Reattach the front of the hub, and then tighten it to secure the dowel rods.

5. Consider the shape and design of blades, and design blades that you think will capture the wind most efficiently. Use the sheets of paper and scissors to create the blades for the turbine.
6. Attach the blades to the dowel rods with tape.
7. Assemble a galvanometer, a device that can measure the relative strength of an electric current. When the galvanometer is aligned with the earth's magnetic field and subjected to a perpendicular current from another source (creating a perpendicular magnetic field), the angle of deflection of

the compass needle from North gives an indication of the strength of the unknown current. The greater the deflection, the stronger the magnetic field and electric current producing it.

    a. Orient the strand of wire in a North-South direction across the face of the compass, with one end of the wire extending slightly from the edge of the compass marked "N" (North). Wrap the wire entirely around the compass from top to bottom (i.e., from North to South) a total of five times. One end of the wire should project North, while the other end projects South.

    b. Tape the wire securely to the compass.

    c. Place the compass on the lab bench, and rotate it until the compass needle points to North. Then, tape the compass to the lab bench.

8. Attach one end of the black alligator clip to the end of the wire at the North end of the compass, and the other end of the black alligator clip to the black wire on the DC motor.

9. Attach one end of the red alligator clip to the end of the wire at the South end of the compass, and the other end of the red alligator clip to the red wire on the DC motor.

10. Use a small fan to move the blades. Vary the speed and direction of the wind and observe the effects on the compass needle.

11. Disassemble the wind turbine and galvanometer. Discard any trash.

12. Answer the questions for the *Wind* activity on Discussion and Conclusion page.

Lab 10

ACTIVITY 4 Battery

Batteries are commonly used to power portable electronic devices such as cell phones, laptops, and e-book readers. They can also be used to supply electricity to larger machines like cars and trucks. There are many different types of batteries, but they all work by converting chemical energy into electrical energy. Some batteries are disposable; others can be recharged numerous times before they will no longer hold a charge and must be disposed of.

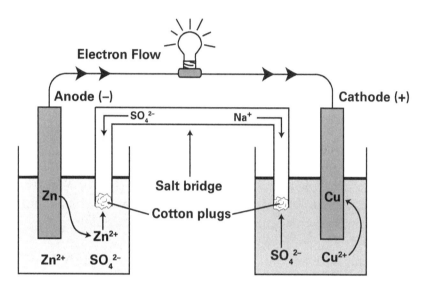

A typical voltaic cell or "wet cell" battery

Batteries are composed of two half-cells separated by a salt bridge. In each half-cell is an electrode bathed in a liquid or paste electrolyte. The anode, the negative electrode, is located at the negative terminal of the battery. At the anode, electrons are lost in an oxidation reaction. The cathode, the positive electrode, is located at the positive terminal of the battery. At the cathode, electrons are gained in a reduction reaction. The cathode accepts the flow of electrons from the anode when the battery is connected in a circuit.

Differences in electrode type can affect the amount of voltage produced. Alkaline batteries, carbon–zinc batteries, and lithium batteries all produce different voltages, which can be used to power devices with different energy requirements.

Energy production from batteries does not release combustion by-products that contribute to climate change, as is the case with energy produced by burning fossil fuels. Nevertheless, there are several significant hazards associated with the electrolytes contained within batteries. Batteries can leak electrolytes, which are often toxic substances such as lead, cadmium, and mercury. It is also possible that batteries can explode when overcharged or misused. Disposal of batteries presents some concerns because toxins can leak from old or damaged batteries and contaminate the environment. Recycling services are available for many types of batteries, but every year thousands of tons of batteries end up in landfills across the United States.

MATERIALS

- D-cell battery
- 9-volt battery
- Battery holder
- Compass
- 2 alligator clips (black and red)
- 2' wire
- Tape

PROCEDURE

1. Assemble a galvanometer, a device that can measure the relative strength of an electric current. When the galvanometer is aligned with the earth's magnetic field and subjected to a perpendicular current from another source (creating a perpendicular magnetic field), the angle of deflection of the compass needle from North gives an indication of the strength of the unknown current. The greater the deflection, the stronger the magnetic field and electric current producing it.
    a. Orient the strand of wire in a North–South direction across the face of the compass, with one end of the wire extending slightly from the edge of the compass marked "N" (North). Wrap the wire entirely around the compass from top to bottom (i.e., from North to South) a total of five times. One end of the wire should project North, while the other end projects South.
    b. Tape the wire securely to the compass.
    c. Place the compass on the lab bench, and rotate it until the compass needle points to North. Then, tape the compass to the lab bench.
2. Attach one end of the black alligator clip to the wire at the North end of the compass.
3. Attach one end of the red alligator clip to the wire at the South end of the compass.
4. Place the D-cell battery in the battery holder.
5. Connect the free end of the black alligator clip to the negative (–) terminal of the battery holder.

6. Connect the free end of the red alligator clip to the positive (+) terminal of the battery holder. Watch the deflection of the compass needle.
7. Disconnect alligator clips from the battery holder.
8. Connect the black alligator clip to the negative terminal of the 9-volt battery.
9. Connect the red alligator clip to the positive terminal of the 9-volt battery. Observe the deflection of the compass needle as the circuit is completed.
10. Disassemble the battery circuit and the galvanometer.
11. Answer the questions for the Battery activity on Discussion and Conclusion page.

Lab 10

Energy Sources. Part 1

Student Name _____   Lab Group _____

TA Name _____   Lab Date/Section _____

DISCUSSION AND CONCLUSION

▶ *Coal Gas*

   1. What happened when the coal shavings were heated?

   2. What happened when the lit toothpick was held up to the pipet tip?

   3. Why was gas produced when the coal was heated?

   4. What do you think are the most significant advantages and disadvantages of using coal and coal gasification as power sources?

▶ *Solar*

5. How did the angle of light hitting the photovoltaic (PV) cell and the cell's distance from the light source affect the current produced by the solar cell? Why might this be so?

6. What do you think are the most significant advantages and disadvantages of using solar cells and solar energy?

▶ *Wind*

7. How many blades did you use on your wind turbine? Describe the shape and size of the blades that you designed for use with the wind turbine.

8. What do you think are the most significant advantages and disadvantages of using turbines and wind energy?

▶ *Battery*

9. Describe the differences you observed between the 9-volt and D-cell batteries when they were connected in a circuit with the galvanometer. Explain the reasons for the differences you observed.

10. What do you think are the most significant advantages and disadvantages of batteries as an energy source?

Lab 11

ENERGY SOURCES.
PART 2

PRE-LAB READING

Textbook: Chapter 19: Fossil Fuels; Chapter 20: Conventional Energy Alternatives; Chapter 21: New Renewable Energy Alternatives.

OBJECTIVES

- Learn about renewable and nonrenewable energy resources.
- Examine different energy technologies, and analyze pros and cons of each.
- Build a waterwheel.
- Observe the properties of natural gas.
- Create a model of how energy can be derived from water beneath the earth's surface.
- Observe how batteries are capable of doing different amounts of work.

ACTIVITIES

1. Natural Gas
2. Geothermal
3. Water
4. Hydrogen

Lab 11

ACTIVITY 1   Natural Gas

Although natural gas is a nonrenewable resource, the United States possesses large natural gas reserves. In the United States and many other countries, methane and other natural gases are commonly used as energy sources, especially for heating purposes. Before it can be used, however, natural gas trapped in rock formations underneath the earth must be located, extracted, and processed. First, experts perform surveys to identify natural gas deposits. Then, drilling is performed and the gas is extracted and carried away through a pipeline. During processing, methane is separated from other gases such as butane and propane. Often, a technique known as hydraulic fracturing or "fracking" is utilized to access the underground gas deposits; deep holes are drilled and massive quantities of pressurized fluid are pumped into the earth, fracturing the rock formations to free the gases. A danger of this process is contamination of groundwater with carcinogens and toxic chemicals. Some studies have found a correlation between hydraulic fracturing near fault lines and the potential for increased geologic instability.

Burning natural gas releases carbon dioxide. Mining and drilling disturb natural areas and can contaminate groundwater. In addition, during the mining and refinement process, methane can leak into the atmosphere. Methane is a powerful greenhouse gas. It is also highly flammable and odorless, a dangerous combination that presents containment and transportation challenges. Therefore, producers add substances called mercaptans to the gas to give it a readily apparent, sharp, and recognizable odor. This easily distinguishable odor is intended to help people detect natural gas leaks and minimize the risk of explosions.

MATERIALS

- Access to a natural gas valve and hose
- Beaker of water
- Beaker
- Liquid dish soap
- Candle
- Plastic spoon
- Matches
- 2 graduated cylinders (10 mL and 100 mL)
- Heat-resistant gloves
- Safety goggles

PROCEDURE

1. Make a soap solution by mixing 5 mL of soap with 100 mL of water in a beaker.
2. Place one end of the tubing on the natural gas outlet. Place the other end of the tubing at the bottom of the beaker of soap solution.
3. Slowly turn on the gas. Allow the natural gas to bubble through the soap solution until bubbles have accumulated on the surface of the solution.
4. Turn off the gas completely, and then carefully remove the hose from the beaker.
5. Use a plastic spoon to remove the bubbles from the top of the solution.
6. While wearing safety goggles and heat-resistant gloves, light a candle.
7. Stand back from the candle, and make sure that nothing hangs near or above the spoon or the candle.
   Then, place the plastic spoon beside the candle and observe the effects.
8. Answer questions for the *Natural Gas* activity on Discussion and Conclusion page.

Lab 11

ACTIVITY 2 Geothermal

Scientists have determined that the core of the earth is as hot or even hotter than the surface of the sun (but not nearly as hot as the sun's core). Heat trapped under the earth's surface can be tapped and used, directly or indirectly, to heat buildings and produce electricity.

Geothermal pumps buried less than 10 feet underground can help both to heat and to cool buildings. The temperature range in the upper 10 feet of earth's crust is 50°F–60°F (10°F–16°C). In the summer, hotter air inside a building can be pumped out and exchanged for air cooled underground. In the winter, warmer air from underground can be pumped into the building to heat it. Some cities have even installed pipelines that channel geothermally warmed water underneath sidewalks and roadways to prevent the accumulation of ice and snow on these surfaces. Geothermal energy can be used to generate electricity in many different ways. In most cases, operators dig a well more than a mile deep, and the hot water or steam brought to the surface is harnessed to drive turbines, generating electricity. In "dry steam" geothermal electric power plants, steam is used directly to power generators; in "flash steam" operations, steam is created in tanks under controlled conditions utilizing differences in pressure and temperature, and then used to turn turbines. After the hot water and steam are used to produce power, they are injected back into the earth.

Noise pollution and the potential for groundwater contamination are concerns associated with geothermal energy production. Another challenge for geothermal energy is that the process can release toxic metals, volatile elements (e.g., mercury, arsenic, and selenium), and noxious gases. For instance, water extracted from underground sources may contain sulfurous gases like hydrogen sulfide, a pollutant that can contribute to acid rain. For this reason, geothermally powered electric plants must use scrubbers and other abatement technologies to remove these potentially harmful substances. Significant successes have been achieved in this area, and geothermal plants have demonstrated a trend of decreasing hydrogen sulfide emissions while increasing energy output. Some of the natural geothermal features in the United States, including the geysers at Yellowstone National Park, are protected from development by the National Park Service.

MATERIALS

- Hot plate
- 200-mL Erlenmeyer flask
- Water
- 1-hole flask stopper with pipet inserted
- Pinwheel
- Plastic pipet
- Heat-resistant gloves
- Safety goggles

PROCEDURE

1. Fill the Erlenmeyer flask with 200 mL of water.
2. Insert the rubber stopper into the top of the flask.
3. Heat the water in the flask on the hot plate. Put on heat-resistant gloves and goggles (if you have not already).
4. When the water reaches a rolling boil, hold the pinwheel over the hole in the rubber stopper. Adjust the distance and angle of the pinwheel in relation to the steam exiting through the pipet. Work with caution; hot water and steam can cause burns. Observe the pinwheel for 2 to 3 minutes.
5. Turn off the heat to the hot plate. Allow the flask to cool to room temperature before attempting to handle it.
6. Answer the questions for the *Geothermal* activity on Discussion and Conclusion page.

Lab 11

ACTIVITY 3   Water

Hydropower, energy produced from moving water, is among the earliest forms of energy used by mankind. In 2011, hydropower was the leading source of renewable energy utilized in the United States. Different methods of generating hydropower have been developed. In run-of-the-river (ROR) hydroelectric systems, water flows through a pipe and the force of the current turns the blades of a turbine to spin a generator, which produces electricity. In pumped storage hydroelectric systems, water is contained in an elevated reservoir behind a dam. Water released from the reservoir flows down through a large pipe called a penstock and the force of the water drives a generator, producing electricity. There are legitimate concerns that the construction of dams on rivers has prevented the upstream migration of fish. To combat this problem, some dams are constructed with features called fish ladders, which allow fish such as salmon and shad to ascend and move past the dam. The operation of a dam can also change the temperature of a river, alter the amount of silt on the streambed above and below the dam, and drastically change other critical environmental conditions.

There is the possibility for developers of hydropower to harness the energy from ocean tides and waves. Ocean water moves toward and away from the shore on a massive scale (tidal action) as well as a local scale (wave action). Tidal and wave action generate enormous amount of force, which can be harnessed to turn turbines. Tidal power already is utilized in France and Canada, and a commercial wave farm is operational in Portugal.

MATERIALS

- Plastic cups
- Stapler
- Plastic plates
- Scissors
- 3' wire
- 500-mL beaker of water
- 1-gallon plastic container
- Thumbtack

PROCEDURE

1. Follow these instructions to construct a waterwheel. Use the image below as a guide.
   a. Use scissors to cut off the rim of two plastic plates, creating two flat discs of the same diameter.
   b. Staple six plastic cups to one plastic plate, so that the open end of the cup faces the outer edge of the plastic plate.
   c. Staple a second plate onto the opposite side of the cups to form a waterwheel.
   d. Use a thumbtack to punch holes through the center of each plate. Then, thread a wire through the center of the wheel.
   e. Wrap the ends of the wire underneath the large plastic container. Twist the ends of the wire until it is taut. The wire should hold the waterwheel stable and keep it from touching the bottom of the plastic container.

2. Slowly pour water from the beaker into the cups at the top of the waterwheel.
3. Pour the water back into the beaker.
4. When you are ready to move to the next station, remove the wire from the waterwheel. Set the wire and plastic container aside to be reused.
5. Answer the questions for the *Water* activity on Discussion and Conclusion page.

Lab 11

ACTIVITY 4   Hydrogen

Composed of only one proton, one neutron, and one electron, hydrogen is the lightest of the elements and the most abundant element in the universe. It easily forms compounds and is an important component of water and many organic compounds. Hydrogen is a promising fuel source because it is efficient and environmentally friendly. Proponents point to these qualities as evidence that hydrogen is ideally suited to replace fossil fuel burning engines in vehicles, with the potential to end people's reliance on gas and oil.

Separating the oxygen atom from the two hydrogen atoms in a water molecule yields hydrogen gas, which is combustible. Burning hydrogen releases heat and creates water. The exhaust from a vehicle that runs on hydrogen gas is water vapor, which poses no threat to people or the environment.

Cost and lack of a hydrogen power infrastructure are some of the challenges associated with using hydrogen as a fuel source. Energy is required to break the bonds in water molecules, and currently much of the energy needed for large-scale production of hydrogen fuel would come from coal or nuclear power plants. Therefore, pollution and use of nonrenewable fuels remain as unresolved issues. Another significant problem is that at present few vehicles run on hydrogen fuel, and consumers would have to purchase new ones or reconfigure existing vehicles. Furthermore, a large-scale move to hydrogen would require the creation of an infrastructure for fuel distribution.

MATERIALS

- Small rectangular container
- 2 graduated cylinders (10 mL and 100 mL)
- Sodium sulfate solution
- Water
- 2 small test tubes
- Matches
- Several toothpicks
- Electrode stand
- 9-volt battery

PROCEDURE

1. Measure 10 mL of sodium sulfate in a graduated cylinder, and then pour it into the rectangular container.

2. Add 140 mL of water to the container.
3. Lay the two test tubes on their sides in the container. Completely fill each one with sodium sulfate solution.
4. Place the electrode stand on its side in the solution so that the electrodes are facing the mouths of the submerged test tubes. Slide a test tube onto each electrode, keeping the tubes submerged and full of solution.
5. Once the test tubes are in position, carefully turn the apparatus upright with the closed ends of the test tubes facing upward. The test tubes should remain filled with solution. If necessary, repeat these steps until you have eliminated any excess air in the test tubes.

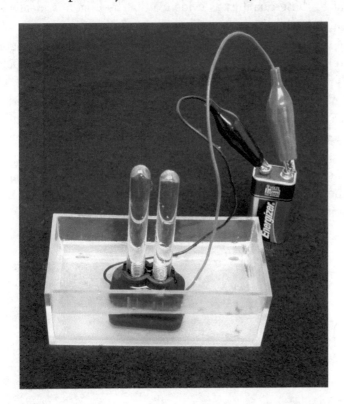

6. Connect the red wire to the positive (+) terminal of the battery. Connect the black wire to the negative (−) terminal of the battery.
7. Observe the gas bubbles being generated at each electrode.
8. When the test tube at the black wire is half to three-quarters full of gas, disconnect the battery from the apparatus.
9. Keeping it inverted (mouth down), remove the test tube containing the most gas from the solution.
10. Use a paper towel to dab the end of the test tube. Keep the tube inverted so that no gas escapes.
11. Have a member of your group use a match to light a toothpick.
12. While holding the test tube at a 45° angle with the mouth down, insert the burning toothpick into the mouth of the test tube. Observe what happens.
13. Remove the test tube containing the lesser amount of gas from the solution, keeping it inverted (mouth down). Use a paper towel to dab the end. Again, keep the test tube inverted so that no gas escapes.
14. Have a member of your group use a match to light a toothpick and then blow out the flame. Immediately insert the toothpick into the tube, and observe what happens.
15. Discard the solution down the drain and rinse all of the remaining equipment with water.
16. Answer the questions for the *Hydrogen* activity on Discussion and Conclusion page.

Lab 11

Student Name _____   Lab Group _____

TA Name _____   Lab Date/Section _____

DISCUSSION AND CONCLUSION

▶ *Natural Gas*

1. In the activity you performed, what do you think would have happened if the natural gas was exposed to a flame without being contained in the soap bubbles? Why?

2. What do you think are the most significant advantages and disadvantages of natural gas power?

▶ *Geothermal*

3. Describe and explain what occurred when the pinwheel was held over the flask of boiling water.

4. What do you think are the most significant advantages and disadvantages of geothermal energy?

▶ *Water*

5. How would you harness the power from the waterwheel to generate electricity?

6. What do you think are the most significant advantages and disadvantages of water power?

▶ *Hydrogen*

7. Sodium sulfate does not form any gas, so what two gases are formed in the test tubes?

8. By observing the gases produced in the test tubes, how could you tell which gas is which?

9. What happened when the lit toothpick was placed in the first test tube? What happened when the extinguished toothpick was placed in the second test tube?

10. What do you think are the most significant advantages and disadvantages of hydrogen fuel?

11. Compare the energy sources that you observed in these laboratory activities, and consider the resources, geography, and climate of your community. Which of these energy sources are the most viable in your community, and why? Which energy sources would you prefer to be used, and why?

Appendix I
BIODEGRADATION PROJECT AND LAB REPORT

PRE-LAB READING

Textbook: Chapter 22: Managing out waste.

ACTIVITIES

1. Project Set Up
2. Project Observation
3. Project Completion
4. Project Lab Report

INTRODUCTION

Human endeavors constantly produce many types of waste, both directly and indirectly. **Municipal solid waste** is the "trash" or "garbage" produced by residences, businesses, and institutions. It consists of a vast array of items and materials including food scraps, newspapers, grass clippings, plastic and glass bottles, and packaging materials. In 2008, every person in the United States produced about 4.5 pounds of waste per day (Quantity of Municipal Solid Waste Generated and Managed, 2010). As the human population has increased, so has the volume of waste generated by our activities. In order to safeguard human health and the environment, we must find effective methods to dispose of waste generated by humans.

Humans have employed a variety of waste disposal techniques through time. When there were a small number of technologically un-advanced humans on the planet there was not much waste generated and that which was generated was organic in composition. Organic waste disposed of on the surface of the ground or in bodies of water decomposed and/or dispersed quickly.

In early human settlements, it was common for people to simply throw their waste out the doors and windows of their homes where it accumulated. It is believed that the city of Athens instituted the first municipal dump when it required that its citizens take their waste at least one mile beyond the city walls for disposal. By the mid-1700s, many American homes had shifted from throwing their waste out their doors and windows to burying it in refuse pits located on their properties. However, in many urban areas of the United States, including Washington, D.C., waste was still dumped in the streets leading to infestations of rodents and cockroaches. By the early 1840s, people started accepting the connection between these unsanitary conditions and disease.

Eventually, the home-based refuse pits were largely replaced by communal "dumps," operated either by a municipality or commercially. Early dumps were simply open trenches or depressions where waste was hauled in and dumped. They were unregulated and lacked human health and environmental safety features. In the absence of these features, toxic materials could migrate off-site in rainwater runoff or in wind-borne dust particles, or end up in the groundwater. By the mid-1940s, sanitary landfills were being used by about 100 cities in the United States.

Sanitary landfills are designed structures that incorporate features that minimize the risk of adverse environmental and human health impacts associated with the land-based disposal of municipal solid waste. They incorporate barriers to minimize the movements of materials and **leachate** (liquid produced when water passes through waste material and becomes contaminated with bacteria or pollutants) off-site or into the groundwater, and the waste is covered daily with a layer of soil. Many sanitary landfills also monitor perimeter wells so that groundwater can be tested to insure that leachate collection systems are functioning and barriers are preventing leaks.

Even with these advances, sanitary landfills still have problems associated with them. One of these problems is that the conditions within the landfill are not conducive to the decomposition of the waste! Decomposition proceeds as a result of the activity of bacteria and fungi, the most effective of which require air and moisture to survive and thrive. The interiors of modern sanitary landfills are relatively dry and airtight.

These conditions were brought to light when archaeologists from the University of Arizona founded the Garbage Project in 1973 and began a series of archaeological digs in a landfill. Included among their finds was a newspaper from 1952 that was still completely intact and readable, and the dried up remains of hot dogs that were 20 years old. Instead of providing a place for **biodegradable** material to biodegrade, the modern sanitary landfill was providing an environment that basically preserved the waste. Even a paper bag, clearly more "biodegradable" than a plastic bag, can take hundreds of years to completely decompose in a modern sanitary landfill. New landfill designs are being studied that would enable the interior of the landfill to receive enough moisture and air to support a thriving population of decomposition microbes.

The microbially aided decomposition process is often initiated by **cellulolytic** bacteria and fungi that are capable of breaking down the cellulose contained in the cell walls of plant waste and converting it into various sugars. These sugars are then fermented by acidogens (acid-generating bacteria) that produce weak acids. The decomposition process is completed when **methanogens** (methane-generating bacteria) convert the weak acids into carbon dioxide and methane. The methane that is produced can be collected and burned to produce electricity.

Sanitary landfills require a large amount of land area. Existing sanitary landfills are filling up and it is becoming more difficult to find suitable, acceptable locations for new sanitary landfills. In the United States, we dispose of about 54% of our municipal solid waste in sanitary landfills.

We can increase the life span of our existing sanitary landfills by minimizing the amount of waste that is put into them. This is called **source reduction** and it is the preferred solid waste management method. Source reduction can be accomplished by adhering to the "**reduce-reuse-recycle**" principle: **reduce** the amount of material you use and/or reduce the amount of products you buy that have excessive packaging; **reuse** items or materials to maximize their life space, thus minimizing the amount of these materials that you must buy, and, ultimately dispose of; and **recycle** materials that are recyclable.

Source reduction can also be accomplished through **composting**, considered to be a type of recycling, which is a controlled biological decomposition of organic matter. Materials such as plant-based food waste and yard waste can be easily composted. Bacteria and fungi play important roles in this decomposition process. Earthworms can also facilitate decomposition. **Worm composting (vermicomposting)** uses worms to convert plant-based food waste and yard waste into a nutrient-rich substance called **vermicompost**, made up of the nutrient-rich earthworm excretions called **castings**.

Another solid waste management technique is **combustion (incineration)** in which the waste is disposed of by burning it at a high temperature. About 14% of the solid waste in the United States is disposed of in this manner. The heat generated by this burning process can be used to generate electricity.

In **activities 1–3 (biodegradation)**, you will compare the biodegradation of several types of materials in sand, dry soil, and moist soil. This will illustrate the differences in the biodegradability of different materials as well as the effect on biodegradation of factors such as soil type and moisture content,

REFERENCES

Botkin, Daniel B. and Edward A. Keller. 2007. Environmental Science: Earth as a Living Planet. John Wiley & Sons, Hoboken, NJ.

Milestones in Garbage. 2006. Municipal Solid Waste. U.S. Environmental Protection Agency. Retrieved October 10, 2007. From http://www.epa.gov/epaoswer/non-hw/muncpl/timeline_alt.htm.

Quantity of Municipal Solid Waste Generated and Managed. 2010. Report on the Environment. U.S. Environmental Protection Agency. Retrieved 23 June 2010. From http://cfpub.epa.gov/eroe/index.cfm?fuseaction=detail.viewInd&Iv=list.listBy Alpha&r=216598&subtop=228.

Rathje, W. L. 2006. The Garbage Project and the Archaeology of Us. Symmetrical Archaeology. Retrieved October 10, 2007. From http://traumwerk.stanford.edu:3455/Symmetry/174.

Ward's Natural Science. 2003. Landfills and the Environment Lab Activity. Ward's Natural Science, Rochester, NY. 31 pp.

Ward's Natural Science. 1992. School Composting Investigations. Ward's Natural Science, Rochester, NY. 8 pp.

APPENDIX I

ACTIVITY 1
Biodegradation Project Set Up

OBJECTIVES

- Understand the role of moisture in the decomposition process.
- To understand the role of soil in the decomposition process.

HYPOTHESIS

- None.

MATERIALS

- Container, plastic, with lid, small, ~12 cm × 12 cm, 3
- Cotton swabs, non-sterile
- Dust pan and brush
- Fiber, natural (cotton string)
- Fiber, synthetic (yarn, or other)
- Gloves, disposable
- Marker, permanent
- Newspaper
- Packing "peanuts," biodegradable
- Packing "peanuts," Styrofoam
- Paper, office, scrap
- Pretzel sticks
- Rubber bands
- Ruler (cm)
- Sand
- Scissors
- Spray bottle, containing tap water
- Stirrers, plastic
- Toothpicks, wooden
- Topsoil

PROCEDURE

1. Obtain three plastic containers with lids (small, ~12 cm × 12 cm) and use a permanent marker to label them as follows:
   - #1-Sand Lab Sec. # Lab Group #
   - #2-Dry topsoil Lab Sec. # Lab Group #
   - #3-Moist topsoil Lab Sec. # Lab Group #
2. Wear disposable gloves when handling soil or any decomposing material.
3. Fill container #1 with sand to within 2 cm of the top and pack it tightly.
4. Fill container #2 and #3 with topsoil to within 2 cm of the top and pack them tightly.
5. Use a spray bottle containing tap water to water the topsoil in container #3 until the soil is moist throughout but not soaking. You do not want to create conditions that are too soggy to promote fungal and aerobic bacterial growth.
6. Observe the original condition of materials to be tested and record your observations in Table A1.1 of the data sheet. Your observations can include descriptions of size, color, shape, intactness, flexibility, etc. The materials to be tested are listed below:
   a. Strip of office paper
   b. Strip of newspaper
   c. Cotton swab
   d. Plastic stirrer piece (or plastic straw piece)
   e. Styrofoam packing "peanut"
   f. Biodegradable packing "peanut"
   g. Synthetic fiber (yarn or nylon thread) piece
   h. Natural fiber (cotton string) piece
   i. Toothpick
   j. Pretzel stick
7. For the two pieces of fiber, you will test their original strength. Refer to the strength testing guide in Table A1.2 of the data sheet. Conduct the strength test by stretching the fiber and determining its strength. Record the fiber strength (under "initial rating") in Table A1.2 of the data sheet.
8. Insert into the sand/soil of each container two pieces of each of the materials to be tested. Be sure to leave a portion of each piece of material exposed above the sand/soil. The exposed portion of the material will serve three purposes: (1) a guide to the original condition of the material, (2) a guide to help you locate it for observations, and (3) a means for you to grip and remove the material from the soil/sand when observations are made.
9. Use your fingers to re-pack the sand/soil around the materials so that they will remain in place.
10. Close and secure the lid on the container.
11. Place your containers in the storage location indicated by the instructor. The containers will be retrieved for observations once in the middle of the semester and again at the end of the semester.
12. **CLEANUP:** When you are finished, make sure the following cleanup steps have been carried out:
    a. Use the dust pan and brush to clean up and discard any spilled soil or sand.
    b. Use paper towels to wipe table clean.
    c. Return all materials to their original locations.
    d. Discard all waste in the appropriate locations.

NOTE: There is no discussion and conclusions section for this activity.

APPENDIX I

ACTIVITY 1
Biodegradation Project Set Up

Student Name: _____ Lab Group: _____

TA: _____ Lab Date/Section: _____

DATA

TABLE A1.1. Initial Observations of Materials to be Degraded, by Treatment
(Sand, Dry Topsoil, Moist Topsoil)

Container #:	1	2	3
Treatment:	Sand	Dry Topsoil	Moist Topsoil
Observation → Material ↓	General appearance and condition: color; texture	General appearance and condition: color; texture	General appearance and condition: color; texture
Office paper			
Newspaper			
Cotton swab			
Plastic stirrer			
Styrofoam "peanut"			
Biodegradable "peanut"			
Synthetic fiber			
Natural fiber			
Toothpick			
Pretzel stick			

APPENDIX I

ACTIVITY
1

Biodegradation Project Set Up

Student Name: _____   Lab Group: _____

TA: _____   Lab Date/Section: _____

DATA

TABLE A1.2. Initial Strength of Natural and Synthetic Fibers

Rating Description	Fiber	Rating
A. Item is in original condition and exhibits original strength. B. Item is showing early signs of decay (i.e., it is discolored, has cracks or small breaks).	Natural	
C. Item shows many signs of decay (color has changed, has large cracks or breaks). D. Item has completely decayed (disintegrated).	Synthetic	

APPENDIX I

ACTIVITY
2 Biodegradation Project Observation

OBJECTIVES

▪ To observe differences in rates of biodegradation (as evidenced by changes in appearance of all materials, and strength of some materials) of different types of materials buried in different types/ and/or condition of soils after a period of 6 to 8 weeks.

HYPOTHESES

▪ Natural/organic materials (paper, newspaper, cotton swab, biodegradable packing peanut, natural fiber, toothpick, pretzel) will show more evidence of biodegradation (evidenced by changes in appearance and/or strength) after 6 to 8 weeks than will synthetic materials (plastic stirrer, Styrofoam packing peanut, synthetic fiber).

▪ Materials buried in moist topsoil will show more evidence of biodegradation (evidenced by changes in appearance and/or strength) after 6 to 8 weeks than will material buried in dry topsoil.

▪ Materials buried in dry topsoil will show more evidence of biodegradation (evidenced by changes in appearance and/or strength) after 6 to 8 weeks than will material buried in sand.

MATERIALS

▪ Magnifying glass
▪ Microscope, dissecting

PROCEDURE

1. Obtain your group's containers from the storage location.
2. Carefully remove from the sand/soil one of the two pieces of one of the non-fiber materials being testing. Gently clean off the sand/soil, and then observe it using either a magnifying glass or a dissecting microscope. Make the same types of observations about the condition of the material as you did when you observed its original condition (characteristics such as size, color, shape, intactness, flexibility, etc.). Record your observations in Table A1.3 of the data sheet. After making your observations on this piece of material, you may discard it according to your instructor's directions.
3. Repeat step 2 until all of the non-fiber materials have been observed.

4. Carefully remove from the sand/soil one of the two pieces of one of the natural fiber material. Gently clean off the sand/soil, and then observe it using either a magnifying glass or a dissecting microscope. Make the same types of observations about the condition of the material as you did when you observed its original condition (characteristics such as size, color, shape, intactness, flexibility, etc.). Record your observations in Table A1.4 of the data sheet.

5. Test the strength of the natural fiber by stretching it and referring to the guide in Table 10.4 of the data sheet. After making your observations on this piece of material, you may discard it according to your instructor's directions.

6. Repeat steps 4 and 5 for one of the two pieces of the synthetic fiber material.

7. Close and secure the lids on the all containers.

8. Place your containers in the storage location indicated by the instructor. The containers will be retrieved for observations again at the end of the semester.

9. **CLEANUP:** When you are finished, make sure the following cleanup steps have been carried out:

    a. Use the dust pan and brush to clean up and discard any spilled soil or sand.

    b. Use paper towels to wipe table clean.

    c. Return all materials to their original locations.

    d. Discard all waste in the appropriate locations.

APPENDIX I

ACTIVITY 2 — Biodegradation Project Observation

Student Name: _____ Lab Group:_____

TA: _____ Lab Date/Section: _____

DATA

TABLE A1.3. Mid-Semester Observations of Materials to be Degraded, by Treatment (Sand, Dry Topsoil, Moist Topsoil)

Container #:	1	2	3
Treatment:	Sand	Dry Topsoil	Moist Topsoil
Observation → / Material ↓	General appearance and condition: color; texture	General appearance and condition: color; texture	General appearance and condition: color; texture
Office paper			
Newspaper			
Cotton swab			
Plastic stirrer			
Twig, small			
Styrofoam "peanut"			
Biodegradable "peanut"			
Synthetic fiber			
Natural fiber			
Toothpick			
Pretzel stick			

APPENDIX I

ACTIVITY 2 Biodegradation Project Observation

Student Name: _____ Lab Group: _____

TA: _____ Lab Date/Section: _____

DATA

TABLE A1.4. Mid-Semester Observation of Strength of Natural and Synthetic Fibers

Rating Description	Fiber	Rating
A. Item is in original condition and exhibits original strength. B. Item is showing early signs of decay (i.e., it is discolored, has cracks or small breaks).	Natural	
C. Item shows many signs of decay (color has changed, has large cracks or breaks). D. Item has completely decayed (disintegrated).	Synthetic	

APPENDIX I

ACTIVITY 2

Biodegradation Project Observation

Student Name: _____   Lab Group:_____

TA: _____   Lab Date/Section: _____

DISCUSSION & CONCLUSIONS

For full credit, questions should be answered thoroughly, in complete sentences, and legibly.

For each hypothesis listed below, state whether or not it was supported by the data collected and explain your response.

1. *Hypothesis:* Natural/organic materials (paper, newspaper, cotton swab, biodegradable packing peanut, natural fiber, toothpick, pretzel) will show more evidence of biodegradation (evidenced by changes in appearance and/or strength) after 6 to 8 weeks than will synthetic materials (plastic stirrer, Styrofoam packing peanut, synthetic fiber)

   _____

   _____

   _____

   _____

   _____

   _____

   _____

   _____

2. *Hypothesis:* Materials buried in moist topsoil will show more evidence of biodegradation (evidenced by changes in appearance and/or strength) after 6 to 8 weeks than will material buried in dry topsoil.

   _____

   _____

   _____

   _____

   _____

   _____

   _____

3. *Hypothesis:* Materials buried in dry topsoil will show more evidence of biodegradation (evidenced by changes in appearance and/or strength) after 6 to 8 weeks than will material buried in sand.

_____

_____

_____

_____

_____

_____

_____

APPENDIX I

ACTIVITY 3 — Biodegradation Project Completion

OBJECTIVES

■ To observe differences in rates of biodegradation (as evidenced by changes in appearance of all materials, and strength of some materials) of different types of materials buried in different types/ and/or condition of soils after a period of 10 to 12 weeks.

HYPOTHESES

■ Natural/organic materials (paper, newspaper, cotton swab, biodegradable packing peanut, natural fiber, toothpick, pretzel) will show more evidence of biodegradation (evidenced by changes in appearance and/or strength) after 10 to 12 weeks than will synthetic materials (plastic stirrer, Styrofoam packing peanut, synthetic fiber)

■ Materials buried in moist topsoil will show more evidence of biodegradation (evidenced by changes in appearance and/or strength) after 10 to 12 weeks than will material buried in dry topsoil.

■ Materials buried in dry topsoil will show more evidence of biodegradation (evidenced by changes in appearance and/or strength) after 10 to 12 weeks than will material buried in sand

MATERIALS

■ Container, for sand collection
■ Container, for topsoil collection
■ Magnifying glass
■ Microscope, dissecting

PROCEDURE

1. Obtain your group's containers from the storage location.
2. Carefully remove from the sand/soil the remaining piece of one of the non-fiber materials being testing. Gently clean off the sand/soil, and then observe it using either a magnifying glass or a dissecting microscope. Make the same types of observations about the condition of the material as you did when you observed its original condition (characteristics such as size, color, shape, intactness, flexibility, etc.). Record your observations in Table A1.5 of the data sheet. After making your observations on this piece of material, you may discard it according to your instructor's directions.
3. Repeat step 2 until all of the non-fiber materials have been observed.

4. Carefully remove from the sand/soil the remaining piece of the natural fiber material. Gently clean off the sand/soil, and then observe it using either a magnifying glass or a dissecting microscope. Make the same types of observations about the condition of the material as you did when you observed its original condition (characteristics such as size, color, shape, intactness, flexibility, etc.). Record your observations in Table A1.6 of the data sheet.

5. Now you will test the strength of the natural fiber by stretching it and referring to the guide in Table A1.6 of the data sheet. After making your observations on this piece of material, you may discard it according to your instructor's directions.

6. Repeat steps 4 and 5 for the remaining piece of the synthetic fiber material.

7. When finished, dispose of all materials according to the directions provided by your instructor.

8. Return sand to the sand collection container.

9. Return top soil to the topsoil collection container.

10. If the small containers used for testing are recyclable - do so. If not, discard according to instructor's directions.

11. **CLEANUP:** When you are finished, make sure the following cleanup steps have been carried out:

    a. Use the dust pan and brush to clean up and discard any spilled soil or sand.

    b. Use paper towels to wipe table clean.

    c. Return all materials to their original locations.

    d. Discard all waste in the appropriate locations.

APPENDIX I

ACTIVITY 3
Biodegradation Project Completion

Student Name: _____   Lab Group:_____

TA: _____   Lab Date/Section: _____

DATA

TABLE A1.5. Final Observations of Materials to be Degraded, by Treatment (Sand, Dry Topsoil, Moist Topsoil)

Container #:	1	2	3
Treatment:	Sand	Dry Topsoil	Moist Topsoil
Observation → Material ↓	General appearance and condition: color; texture	General appearance and condition: color; texture	General appearance and condition: color; texture
Office paper			
Newspaper			
Cotton swab			
Plastic stirrer			
Twig, small			
Styrofoam "peanut"			
Biodegradable "peanut"			
Synthetic fiber			
Natural fiber			
Toothpick			
Pretzel stick			

APPENDIX I

ACTIVITY 3
Biodegradation Project Completion

Student Name: _____   Lab Group:_____

TA: _____   Lab Date/Section: _____

DATA

TABLE A1.6. Final Observation of Strength of Natural and Synthetic Fibers

Rating Description	Fiber	Rating
A. Item is in original condition and exhibits original strength. B. Item is showing early signs of decay (i.e., it is discolored, has cracks or small breaks). C. Item shows many signs of decay (color has changed, has large cracks or breaks). D. Item has completely decayed (disintegrated).	Natural	
	Synthetic	

APPENDIX I

ACTIVITY 3

Biodegradation Project Completion

Student Name: _____   Lab Group:_____

TA: _____   Lab Date/Section: _____

DISCUSSION & CONCLUSIONS

For full credit, questions should be answered thoroughly, in complete sentences, and legibly.

1. You set up three environments (sand, dry soil, wet soil) in which to assess biodegradation of many types of materials over time. In general, *describe* the rate and extent of degradation of the various materials in each of the three environments.

   _____

   _____

   _____

   _____

   _____

   _____

2. *Provide* possible *explanations* for the general degradation trends observed in each environment, as described in #1 above.

   _____

   _____

   _____

3. *Which* materials, if any, were *completely degraded* by the end of this activity? *Was* there any difference in the degradation of these materials in the three different environments?

   _____

   _____

   _____

4. *Provide* possible *explanations* for your answers in #3 above.

   _____

   _____

   _____

5. *Which* materials, if any, were *partially degraded* by the end of this activity? *Was* there any difference in the degradation of these materials in the three different environments?

_____

_____

_____

6. *Provide* possible *explanations* for your answers in #5 above.

_____

_____

_____

7. *Which* materials, if any, showed *no signs of degradation* by the end of this activity? *Was* there any difference in the degradation of these materials in the three different environments?

_____

_____

_____

8. *Provide* possible *explanations* for your answers in #7 above.

_____

_____

_____

9. Based on the results of this activity, *which* material, soil or sand, should be used in a landfill?

_____

_____

_____

10. Based on the results of this activity, *is it important* for water to be allowed to penetrate the landfill? *Why or why not?*

_____

_____

_____

11. Based on what you have learned from this activity in terms of the different rates of degradation of various materials, *what* can manufacturers and consumers do to help slow the rate at which landfills are being filled up?

_____

_____

_____

12. *Identify* possible sources of *error* in the execution of this activity.

_____

_____

_____

For each hypothesis listed below, state whether or not it was supported by the data collected and explain your response.

13. *Hypothesis:* Natural/organic materials (paper, newspaper, cotton swab, biodegradable packing peanut, natural fiber, toothpick, pretzel) will show more evidence of biodegradation (evidenced by changes in appearance and/or strength) after 10 to 12 weeks than will synthetic materials (plastic stirrer, Styrofoam packing peanut, synthetic fiber).

_____

_____

_____

_____

_____

_____

14. *Hypothesis:* Materials buried in moist topsoil will show more evidence of biodegradation (evidenced by changes in appearance and/or strength) after 10 to 12 weeks than will material buried in dry topsoil.

_____

_____

_____

_____

_____

15. *Hypothesis:* Materials buried in dry topsoil will show more evidence of biodegradation (evidenced by changes in appearance and/or strength) after 10 to 12 weeks than will material buried in sand.

_____

_____

_____

_____

_____

APPENDIX I

ACTIVITY 4 Biodegradation Project Lab Report

Any scientist who tests a hypothesis and performs an experimental investigation has to document 'the steps and details of the experiment, summarize and analyze the results, and make a conclusion about the results supporting or not supporting the original hypothesis. In order to share these results with the scientific community, the scientists have to write a report following an established scientific format and style. This report, called a scientific paper, can be presented orally at a scientific meeting and/or published in one of the specialized scientific journals.

You will have to write a Lab Report based on the hypotheses and results of your Biodegradation Project following an established scientific format. Use the guideline provided bellow. The entire report should be **6–7 pages long**, no more and no less. Although the Biodegradation Project was a group experiment, each student must prepare and write **his/her own** lab report.

INSTRUCTION FOR WRITTEN LAB REPORT

▶ General Format

The lab report should be written in the passive voice, not in first person. For example, instead of saying "I added soil to the column . . . ," you would say "Soil was added to the column . . . ."

There is a 7-page length limit for the lab report. The goal is to say everything that must and should be said as concisely. A 12 point font should be used along with margins that are 1" on the top and bottom of the page and 1.25" on the left and right sides of the page.

The lab report must contain a cover page that lists the title of the lab, centered left to right and top to bottom. Centered at the bottom of the cover page, each on its own line, will be your name, the date of submission, your lab instructor's name, and your lab section number.

The next page will contain the abstract, which should be single-spaced, with the section heading centered above it.

The remainder of the lab report will begin on the next page. You should not start a new page for each new section, except for the literature cited section, which should begin on a page of its own. The section name should appear, centered on the page, prior to the start of each section.

The introduction, methods, results, and discussion sections should be double spaced. The citation on the literature cited page should be arranged in alphabetical order by author, singled-spaced within a citation, double-spaced between citations, with a hanging indent of 0.5."

The lab reports must be submitted electronically as an MS Word document and as a hard copy to your TA.

The written lab report must contain the following items/sections:

- Title
- Abstract
- Introduction
- Methods
- Results

- Discussion
- Literature Cited

You can go to http://www.wisc.edu/writing/Handbook/S cienceReport.html for a good, concise guide to writing lab reports.

Each of these items/sections will be discussed below and an example of most will be provided.

► Title

The lab report must have a title and the title must be descriptive of the experiment. It is *not* appropriate to title a lab report "Lab Report" or "Winogradsky Column Experiment"! The title should appear on a cover page, mid-way down and centered. At the bottom of the cover page, centered, you should place your name, your lab instructor's name, your lab section number, and the date of submission.

In italics below is a sample title:

The Effects of Aerially Applied Diflubenzuron on Decomposition Rate
and Litter Arthropods in Prince William County, Virginia

► Abstract

The first section of the written lab report should be the **abstract**. It can be challenging to write a thorough but concise abstract. The length of the abstract should be no more than one page and ideally should be shorter than one page.

Although it will appear first, it is sometimes suggested that it be written last. The abstract is a miniature version of the lab report. Its purpose is to provide potential readers of the report (or article in a scientific journal) enough information about the experiment to decide whether or not they should read the entire report. Obviously, the material in the abstract should be presented in the same order in which it will be encountered in the lab report.

The abstract should be on a page by itself, following the title page and before the next page, which will begin with the introduction. An abstract sample is not provided here.

► Introduction

The **Introduction** section of the lab report "introduces" the reader to the experiment. The problem or issue being studied must be described. Background information on this problem/issue must be provided. This requires doing some research into the problem/issue and providing a summary of the key information. This background information should also familiarize the reader with the topic. The source of the background information must be cited in the body of the introduction section. All sources cited must be compiled in the "Literature Cited" section. For this report, students must cite literature from three sources in their introduction section.

The introduction must include a brief description of the experiment and a statement of the **hypotheses**. In order for your statement about your hypothesis to be complete, it needs to include a statement about an **observation** that you expect to make, as well as an **inference** explaining why you think it will occur.

The section in italics below is an incomplete excerpt from an introduction section:

Introduction

The purpose of the research described here was to evaluate the effects of repeated applications of diflubenzuron on the decomposition rate of leaf litter and on non-target arthropds residing in the leaf litter.

Since litter arthropods play an important role in the decomposition of leaf litter, significant changes in litter arthropod populations as a result of diflubenzuron applications should be reflected in changes in leaf litter decomposition rates.

▶ Methods

The **Methods** section of the lab report should contain an explanation, in paragraph format, of the procedures that were carried out to test the hypotheses in the experiment. *Do not* provide a bulleted list of the materials used and/or a bulleted or numbered list of the steps taken to carry out the experiment.

Don't be mislead or confused by the "materials" and "procedure" sections of most of the lab exercises in *this* lab manual. The "materials" needed for each exercise are provided in a list for ease in setting up the exercise and obtaining the necessary supplies and equipment. Since most of the exercises in this book are not being designed by the student, the "procedure" section presents the steps necessary for carrying out the exercise in a numbered list so that the majority of students using this manual will be able to carry out the unfamiliar activities.

The person reading your lab report should be able to carry out the experiment exactly the way you did based on your written methods. If the reader is confused, then you have not written the methods properly or clearly.

The reader also needs to know why you carried out the steps and processes. For example, a study that investigated the effects of a medication on liver function might have involved 100 people. The report might have stated, "The liver function of 100 people was tested after they had taken the medication for four months." But for a reader or scientist wanting to replicate the experiment, it would be important to know at least the ages and genders of the human subjects. Obviously, using 100 10-year-old males in such a study would produce very different results than using 100 80-year-old males or 100 50-year-old females. It would be clearer for the lab report to include a more descriptive statement that also included a reason, such as "This study used 50 men and 50 women, all of which were 50 years old, so that all subjects were the some age and so that both genders were equally represented."

As you describe your procedures, make sure you include all of the materials you used and the quantity of each material used. Provide additional detail about the materials when it is necessary and pertinent. In this Winogradsky column experiment, plastic tubes were used. For reader or someone wanting to replicate the experiment it would be important to know that the plastic tubes were clear (reason: so that changes in the microbial growth could be observed) and the size of tube used. It is also important to indicate in this section what data will be collected, when it will be collected, and how it will be analyzed.

The section in italics below is an excerpt from a methods section.

Methods

Three study areas were selected from mixed oak-hickory deciduous forests. Two study areas in western Prince William County had been treated for gypsy moth with diflubenzuron three years and five years, respectively. The third study area, in eastern Fauquier County, served as the control because it had never been treated with diflubenzuron. Table 3 summarizes the diflubenzuron treatment histories of the study areas.

Leaf litter was collected from each of the 10 sample sites within each of the three study areas on each sampling date. The litter sampling location at each site was varied in a random manner from one sampling date to the next. When a litter sample was taken, the metal stake, serving as a permanent marker for the plot, was located Random compass coordinates and a random number between 1 and 5 were generated using a random numbers table. The litter sample location was selected by moving away from the permanent marker a distance of 1 m to 5 m in a compass direction, as indicated by the random numbers table.

All leaf litter, including the first 10 mm of soil, was gathered from a 0.25 m² area and placed in a (15 cm by 46 cm) plastic sample bag until the bag was approximately 3/4 full. To allow ventilation, the bags were loosely sealed by twisting the open end and securing with a metal twist tie. Each sample bag was labeled to identify sample location and date.

▶ Results

The **results** section consists of a written portion as well as either a graph or a table that summarizes the data collected in the experiment.

In the written portion you should state the summary values for key data (such as mean, mode, median, range). You do not need to write out every bit of data collected. *You should not make any interpretation of your data in this section!*

If there is a trend in the data it should be described. A trend is a verbal summary of a relationship between two variables. The following are examples of trend descriptions:

- *The average number of trees chewed on by beavers increased the closer the trees were to the water.*
- *In our sample of 50 people, those that had children all indicated more headaches than those without children.*

All of your data must summarized in either table or graph form. It is not necessary to present the same data in both a graph and a table. You should choose and use which ever of the two provides the best summary/illustration of your results. Each graph must be labeled with the word "Figure" followed by a number. The first graph would be labeled "Figure 1," the second graph would be "Figure 2," etc. Each table is labeled with the word "Table" followed by a number. The first table would be labeled "Table 1," the second table would be "Table 2," etc. Since you want the reader to look at your graphs/tables, you must alert the reader to their presence in the lab report by referencing them in the results section. Every graph and table must be referenced individually in this section. An example of a reference follows:

- *See Figure 1 for all tree data.*

All graphs and tables must have a thoroughly descriptive title in addition to the "Figure" or "Table" label. The title must include both the dependent and the independent variables and should immediately follow the figure/table label. Examples of graph and table labels and titles follow:

- *Figure 1. Mean distance from the water (in meters) of chewed and not-chewed trees selected by beavers.*
- *Table 1. Selection frequency of trees by beavers based on tree species.*

Both the *x*- and *y*-axis of graphs must be labeled with both the name of the parameter being graphed on that axis *and* the units that were used. Also, you must make sure that you place the correct variable on the appropriate axis. The dependent variable goes on the *y*-axis and the independent variable goes on the *x*-axis.

The section in italics below is an excerpt from a results section.

Results

Isopoda were more abundant at the control site than at either of the spray sites on 3 of the 9 sampling dates. More Isopoda were collected at the 3-year spray sites than at the control or 5-year spray sites on 4 dates. The largest Isopoda collection was 18 individuals on 18 May 1992. The smallest collection was 2 individuals on 6 May 1993.

▶ Discussion

The discussion section must include a restatement of your hypotheses. And remember that the hypotheses must include an observation and an inference.

In the discussion section you must state your conclusion as to whether or not the data supported each of your hypotheses. You should never use the words "proved" or "disproved." Data collected in an experiment can only support or not support a hypothesis. The data, in summary form, that led to your conclusion must also be restated in the discussion section.

If your hypothesis was not supported, you must attempt in this section to explain why by providing at least one alternative explanation for your results. Was there an error in the experiment? Did you not control all the variables?

If your hypothesis was supported, you must state at least one alternative explanation for why your hypothesis was supported. In other words, is there a variable that may not have been controlled for that could be the reason the data supported the hypothesis?

The discussion section is where you should pose new questions that have arisen as a result of this experiment and to formulate new hypotheses.

This is the only section of the lab report in which you interpret your data. In every other section you are only presenting facts. In this section you should explore and discuss your observations, data, and conclusions.

The section in italics below is an excerpt from a discussion section.

Discussion

The expected result of this study was that, since diflubenzuron adversely affects arthropods, arthropod abundance in the 3-year and the 5-year spray sites would be reduced relative to the control site. The results of this study, however, do not demonstrate such an affect. In fact, the results support an opposite scenario. a greater abundance of litter arthropods in the spray sites relative to the control site. Table 33 summarizes the dates on which the 3-year or the 5-year spray sites had a greater arthropod abundance than the control site, presented by taxon and date. Litter arthropod abundance was greater in the spray sites than in the control site in 53.7% of the comparisons, and significantly greater in 17.5% of those comparisons.

Gypsy moth infested areas that are not treated to suppress the gypsy moths are likely to experience episodes of canopy defoliation (USDA 1995). Such defoliation can alter the microhabitat of the forest litter by altering its moisture content and temperature (Schowalter and Sabin 1991, Neumann 1991). Alteration of the forest litter microhabitat can, in turn, lead to changes in litter arthropod abundance (Bade, jo and Straalen 1993, Schowalter and Sabin 1991, gird and Chatarpaul 1988, MacKay et al. 1986, Vats and Honda 1988). In this study, a possible explanation for the greater abundance of litter arthropods in the spray sites than in the control site is that the effects of gypsy moth infestation on the canopy in the control site resulted in an alteration of the forest litter microhabitat that did not favor litter arthropods (such as a decrease in moisture content and an increase in temperature) In theory, the repeated treatment with diflubenzuron of the gypsy moth infestations in the 3-year and 5-year spray sites preserved the canopy, which in turn preserved the litter microhabitat that favored litter arthropods (such as higher moisture content and lower temperature) (USDA 1995).

▶ Literature Cited

You are required to cite literature from three sources in the introduction section of your lab report. You must indicate in the body of the lab report when you are citing literature by giving the author's last name followed by the publication date or copyright date (not separated by commas) in parentheses immediately after stating the pertinent information. For example:

Spider populations are adversely affected by multiple-year application of diflubenzuron (Largen 2000).

In the "literature cited" section at the end of the lab report you must list, in alphabetical order by author, the works that you actually cited in the body of the report. Examples of the format for listing literature cited follow:

Uetz, 6. W 1974. A method or measuring habitat space in studies of hardwood forest litter arthropods Environ. Entomol 3: 313–315.

Borror, D. J., C. A. Triplehorn, and N. F. Johnson. 1989. An introduction to the study of insects, 6th edition. Saunders College Publishing of Holt, Rinehart and Winston, Inc., Orlando, FL. 875 pp.

Appendix II

BLANK DATA SHEETS

Lab/Activity Title _____

Student Name: _____ Lab Group: _____

TA: _____ Lab Date/Section: _____